BASICS OF
MATLAB®
and Beyond

BASICS OF
MATLAB®
and Beyond

Andrew Knight

CHAPMAN & HALL/CRC

Boca Raton   London   New York   Washington, D.C.

Library of Congress Cataloging-in-Publication Data

Knight, Andrew (Andrew James), 1961—
    Basics of MATLAB and beyond / Andrew Knight.
      p. cm.
    Includes index. (alk. paper)
    ISBN0-8493-2039-9
    1. Engineering mathematics—Data processing. 2. MATLAB. I. Title.
TA345.K63   1999
620′.001′5118—dc21
                                         99-31210
                                             CIP

Visit the CRC Press Web site at www.crcpress.com

Preface

This book arose from notes written for MATLAB® training courses run within the Australian Defence Science and Technology Organisation. The book is in two parts. Each part was originally a two-day course, designed assuming that students were seated at a computer with MATLAB running.

Part 1 is an introductory course suitable for those with no experience at all with MATLAB. It is written in a self contained way; if you go through the notes, all the new commands and ideas are explained as they are introduced.

Part 2 is a more advanced course suitable for those who are already familiar with the basics of MATLAB. It covers a variety of topics, some of which you may not be interested in; if so, you should be able to skip that section without detriment to other sections.

You can get the m-files that accompany this book from the "Download" section of the CRC Press web site (www.crcpress.com). The files are available in zip or gzipped tar format, and can be extracted using WinZip on a PC, or by using gunzip and tar on UNIX. You will need to put them in a directory where MATLAB will be able to find them. You can either use the cd command to move MATLAB's working directory to the directory you extract the files to, or add that directory to MATLAB's search path. (You can display MATLAB's current working directory by

MATLAB is a registered trademark
of The MathWorks, Inc. For product
information, please contact:

The MathWorks, Inc.
24 Prime Park Way
Natick, MA 01760-1500 USA

Tel: 508-647-7000
Fax: 508-647-7107
E-mail: info@mathworks.com
Web: www.mathworks.com

typing `pwd` (print working directory) in the command window.) On a PC or Macintosh, you can add directories to MATLAB's path by clicking on the path browser button at the top of the MATLAB command window (it is the button with two folders on it to the left of the question mark button). In the path browser, select the menu "Path→Add to path", then select the directory containing the extracted files using the browse button (on PCs it is the one with three dots on it), then check the "add to back" option before pressing "OK". Then click "File→Save Path" before you exit the path browser. If you are using another platform you can use the `path` command from within MATLAB (type `help path` for instructions). You can install this path each time you start MATLAB by putting an appropriate path command in a file called startup.m in a directory called matlab situated immediately below your home directory.

Many of the graphical examples in this book assume that the figure window is empty. To ensure an empty figure window issue the command:

`clf`

which stands for "clear figure". If you find that the figure window is obscured by your command window, try shrinking both windows. Or you can type:

`shg`

(show graphic) to bring the graphics window to the front. The companion software implements an even shorter abbreviation; type

`s`

to bring the graphics window to the front.

If, on a PC or Macintosh, the figure window is at the front of the screen, or if it has the current focus, just start typing and MATLAB will switch to the command window and accept your typing.

Words appearing in this book in typewriter font, for example, `type`, represent MATLAB commands that you can type in, or output produced by MATLAB.

Andrew Knight

About the Author

The author completed a Ph.D. in plasma physics at the Flinders University of South Australia in the days before MATLAB. Consequently, he knows how much time can be saved when you don't have to write your own matrix inversion or polynomial evaluation routines in FORTRAN. His first exposure to MATLAB was at the Centre for Plasma Physics Research at the Swiss Federal Institute of Technology (Ecole Polytechnique Fédérale) in Lausanne, Switzerland, where he continued his research in plasma physics. On his return to Australia to take up a position with the Maritime Operations Division of the Defence Science and Technology Organisation, he was given responsibility for research in the flow noise problem of towed sonar arrays. His current research interests include sonar signal processing and information displays. He has been largely responsible for the growth in the use of MATLAB in his division, and has conducted training courses in MATLAB.

Contents

Basics of MATLAB

1   First Steps in MATLAB

1.1   Starting MATLAB

MATLAB is a software package that lets you do mathematics and computation, analyse data, develop algorithms, do simulation and modelling, and produce graphical displays and graphical user interfaces.

To run MATLAB on a PC double-click on the MATLAB icon. To run MATLAB on a UNIX system, type `matlab` at the prompt.

You get MATLAB to do things for you by typing in commands. MATLAB prompts you with two greater-than signs (>>) when it is ready to accept a command from you.

To end a MATLAB session type `quit` or `exit` at the MATLAB prompt.

You can type `help` at the MATLAB prompt, or pull down the Help menu on a PC.

When starting MATLAB you should see a message:

```
To get started, type one of these commands: helpwin,
helpdesk, or demo
>>
```

The various forms of help available are

`helpwin`	Opens a MATLAB help GUI
`helpdesk`	Opens a hypertext help browser
`demo`	Starts the MATLAB demonstration

The complete documentation for MATLAB can be accessed from the hypertext helpdesk. For example, clicking the link <u>Full Documentation</u>

<u>Set</u> → Getting Started with MATLAB will download a portable document format (PDF) version of the *Getting Started with MATLAB* manual.

You can learn how to use any MATLAB command by typing `help` followed by the name of the command, for example, `help sin`.

You can also use the `lookfor` command, which searches the help entries for all MATLAB commands for a particular word. For example, if you want to know which MATLAB functions to use for spectral analysis, you could type `lookfor spectrum`. MATLAB responds with the names of the commands that have the searched word in the first line of the help entry. You can search the entire help entry for all MATLAB commands by typing `lookfor -all` *keyword*.

1.2 First Steps

To get MATLAB to work out $1 + 1$, type the following at the prompt:

```
1+1
```

MATLAB responds with

```
ans  =
     2
```

The answer to the typed command is given the name `ans`. In fact `ans` is now a variable that you can use again. For example you can type

```
ans*ans
```

to check that $2 \times 2 = 4$:

```
ans*ans
ans  =
     4
```

MATLAB has updated the value of `ans` to be 4.

The spacing of operators in formulas does not matter. The following formulas both give the same answer:

```
1+3 * 2-1 / 2*4
1 + 3 * 2 - 1 / 2 * 4
```

The order of operations is made clearer to readers of your MATLAB code if you type carefully:

```
1 + 3*2 - (1/2)*4
```

1.3 Matrices

The basic object that MATLAB deals with is a matrix. A matrix is an array of numbers. For example the following are matrices:

$$\begin{pmatrix} 12 & 3 & 9 \\ -1200 & 0 & 1e6 \\ 0.1 & \mathrm{pi} & 1/3 \end{pmatrix}, \begin{pmatrix} 1 & 2 & 3 & 4 & 5 \end{pmatrix}, \begin{pmatrix} i \\ -i \\ i \\ -i \end{pmatrix}, 42.$$

The size of a matrix is the number of rows by the number of columns. The first matrix is a 3×3 matrix. The (2,3)-element is one million—1e6 stands for 1×10^6—and the (3,2)-element is pi = $\pi = 3.14159\ldots$. The second matrix is a row-vector, the third matrix is a column-vector containing the number i, which is a pre-defined MATLAB variable equal to the square root of -1. The last matrix is a 1×1 matrix, also called a scalar.

1.4 Variables

Variables in MATLAB are named objects that are assigned using the equals sign = . They are limited to 31 characters and can contain upper and lowercase letters, any number of '_' characters, and numerals. They may not start with a numeral. MATLAB is case sensitive: A and a are different variables. The following are valid MATLAB variable assignments:

```
a = 1
speed = 1500
BeamFormerOutput_Type1 = v*Q*v'
name = 'John Smith'
```

These are invalid assignments:

```
2for1 = 'yes'
first one = 1
```

To assign a variable without getting an echo from MATLAB end the assignment with a semi-colon ;. Try typing the following:

```
a = 2
b = 3;
c = a+b;
d = c/2;
d
who
whos
clear
who
```

1.5   The Colon Operator

To generate a vector of equally-spaced elements MATLAB provides the colon operator. Try the following commands:

```
1:5
0:2:10
0:.1:2*pi
```

The syntax *x:y* means roughly "generate the ordered set of numbers from *x* to *y* with increment 1 between them." The syntax *x:d:y* means roughly "generate the ordered set of numbers from *x* to *y* with increment *d* between them."

1.6   Linspace

To generate a vector of evenly spaced points between two end points, you can use the function linspace(*start,stop,npoints*):

```
>> x = linspace(0,1,10)
x   =
  Columns 1 through 7
        0   0.1111   0.2222   0.3333   0.4444   0.5556   0.6667
  Columns 8 through 10
  0.7778   0.8889   1.0000
```

generates 10 evenly spaced points from 0 to 1. Typing linspace(*start, stop*) will generate a vector of 100 points.

1.7   Plotting Vectors

Whereas other computer languages, such as FORTRAN, work on numbers one at a time, an advantage of MATLAB is that it handles the matrix as a single unit. Let us consider an example that shows why this is useful. Imagine you want to plot the function $y = \sin x$ for x between 0 and 2π. A FORTRAN code to do this might look like this:

```
DIMENSION X(100),Y(100)
PI = 4*ATAN(1)
DO 100 I = 1,100
    X(I) = 2*PI*I/100
    Y(I) = SIN(X(I))
100 CONTINUE
PLOT(X,Y)
```

Here we assume that we have access to a FORTRAN plotting package in which PLOT(X,Y) makes sense. In MATLAB we can get our plot by typing:

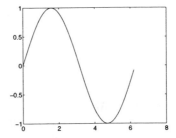

```
x = 0:.1:2*pi;
y = sin(x);
plot(x,y)
```

The first line uses the colon operator to generate a vector x of numbers running between 0 and 2π with increment 0.1. The second line calculates the sine of this array of numbers, and calls the result y. The third line produces a plot of y against x. Go ahead and produce the plot. You should get a separate window displaying this plot. We have done in three lines of MATLAB what it took us seven lines to do using the FORTRAN program above.

2   Typing into MATLAB

2.1   Command Line Editing

If you make a mistake when entering a MATLAB command, you do not have to type the whole line again. The arrow keys can be used to save much typing:

↑	ctrl-p	Recall previous line
↓	ctrl-n	Recall next line
←	ctrl-b	Move back one character
→	ctrl-f	Move forward one character
ctrl-→	ctrl-r	Move right one word
ctrl-←	ctrl-l	Move left one word
home	ctrl-a	Move to beginning of line
end	ctrl-e	Move to end of line
esc	ctrl-u	Clear line
del	ctrl-d	Delete character at cursor
backspace	ctrl-h	Delete character before cursor
	ctrl-k	Delete (kill) to end of line

If you finish editing in the middle of a line, you do not have to put the cursor at the end of the line before pressing the return key; you can press return when the cursor is anywhere on the command line.

2.2   Smart Recall

Repeated use of the ↑ key recalls earlier commands. If you type the first few characters of a previous command and then press the ↑ key

MATLAB will recall the last command that began with those characters.
Subsequent use of ↑ will recall earlier commands that began with those
characters.

2.3   Long Lines

If you want to type a MATLAB command that is too long to fit on one
line, you can continue on to the next by ending with a space followed by
three full stops. For example, to type an expression with long variable
names:

```
Final_Answer = BigMatrix(row_indices,column_indices) + ...
               Another_vector*SomethingElse;
```

Or to define a long text string:

```
Mission = ['DSTO''s objective is to give advice that' ...
'is professional, impartial and informed on the' ...
'application of science and technology that is best' ...
'suited to Australia''s defence and security needs.'];
```

2.4   Copying and Pasting

Your windowing system's copy and paste facility can be used to enter
text into the MATLAB command line. For example all of MATLAB's built-
in commands have some helpful text that can by accessed by typing `help`
followed by the name of the command. Try typing `help contour` into
MATLAB and you will see a description of how to create a contour plot.
At the end of the help message is an example. You can use the mouse
to select the example text and paste it into the command line. Try it
now and you should see a contour plot appear in the figure window.

3   Matrices

3.1   Typing Matrices

To type a matrix into MATLAB you must

- begin with a square bracket [

- separate elements in a row with commas or spaces

- use a semicolon ; to separate rows

- end the matrix with another square bracket].

For example type:

```
a = [1 2 3;4 5 6;7 8 9]
```

MATLAB responds with

```
a   =
    1       2       3
    4       5       6
    7       8       9
```

3.2   Concatenating Matrices

Matrices can be made up of submatrices: Try this:

```
>> b = [a 10*a;-a [1 0 0;0 1 0;0 0 1]]
b   =
    1       2       3      10      20      30
    4       5       6      40      50      60
    7       8       9      70      80      90
   -1      -2      -3       1       0       0
   -4      -5      -6       0       1       0
   -7      -8      -9       0       0       1
```

The `repmat` function can be used to replicate a matrix:

```
>> a = [1 2; 3 4]
a   =
    1       2
    3       4
>> repmat(a,2,3)
ans   =
    1       2       1       2       1       2
    3       4       3       4       3       4
    1       2       1       2       1       2
    3       4       3       4       3       4
```

3.3   Useful Matrix Generators

MATLAB provides four easy ways to generate certain simple matrices. These are

zeros	a matrix filled with zeros
ones	a matrix filled with ones
rand	a matrix with uniformly distributed random elements
randn	a matrix with normally distributed random elements
eye	identity matrix

To tell MATLAB how big these matrices should be you give the functions the number of rows and columns. For example:

```
>> a = zeros(2,3)
a   =
     0     0     0
     0     0     0

>> b = ones(2,2)/2
b   =
    0.5000      0.5000
    0.5000      0.5000

>> u = rand(1,5)
u   =
    0.9218      0.7382      0.1763      0.4057      0.9355

>> n = randn(5,5)
n   =
   -0.4326      1.1909     -0.1867      0.1139      0.2944
   -1.6656      1.1892      0.7258      1.0668     -1.3362
    0.1253     -0.0376     -0.5883      0.0593      0.7143
    0.2877      0.3273      2.1832     -0.0956      1.6236
   -1.1465      0.1746     -0.1364     -0.8323     -0.6918

>> eye(3)
ans   =
     1     0     0
     0     1     0
     0     0     1
```

3.4   Subscripting

Individual elements in a matrix are denoted by a row index and a column index. To pick out the third element of the vector u type:

```
>> u(3)
ans   =
    0.1763
```

You can use the vector [1 2 3] as an index to u. To pick the first three elements of u type

```
>> u([1 2 3])
ans   =
    0.9218      0.7382      0.1763
```

Remembering what the colon operator does, you can abbreviate this to

```
>> u(1:3)
ans =
    0.9218    0.7382    0.1763
```

You can also use a variable as a subscript:

```
>> i = 1:3;
>> u(i)
ans =
    0.9218    0.7382    0.1763
```

Two dimensional matrices are indexed the same way, only you have to provide two indices:

```
>> a = [1 2 3;4 5 6;7 8 9]
a =
    1    2    3
    4    5    6
    7    8    9
>> a(3,2)
ans =
    8
>> a(2:3,3)
ans =
    6
    9
>> a(2,:)
ans =
    4    5    6
>> a(:,3)
ans =
    3
    6
    9
```

The last two examples use the colon symbol as an index, which MATLAB interprets as the entire row or column.

If a matrix is addressed using a single index, MATLAB counts the index down successive columns:

```
>> a(4)
ans =
    2
>> a(8)
ans =
    6
```

> **Exercise 1** *Do you understand the following result? (Answer on page 183.)*

```
>> [a a(a)]
ans  =
        1     2     3     1     4     7
        4     5     6     2     5     8
        7     8     9     3     6     9
```

The colon symbol can be used as a single index to a matrix. Continuing the previous example, if you type

```
a(:)
```

MATLAB interprets this as the columns of the a-matrix successively strung out in a single long column:

```
>> a(:)
ans  =
        1
        4
        7
        2
        5
        8
        3
        6
        9
```

3.5   End as a subscript

To access the last element of a matrix along a given dimension, use end as a subscript (MATLAB version 5 or later). This allows you to go to the final element without knowing in advance how big the matrix is. For example:

```
>> q = 4:10
q  =
        4     5     6     7     8     9    10
>> q(end)
ans  =
       10
>> q(end-4:end)
ans  =
        6     7     8     9    10
>> q(end-2:end)
ans  =
        8     9    10
```

This technique works for two-dimensional matrices as well:

```
>> q = [spiral(3) [10;20;30]]
q =
      7      8      9     10
      6      1      2     20
      5      4      3     30
>> q(end,end)
ans =
     30

>> q(2,end-1:end)
ans =
      2     20

>> q(end-2:end,end-1:end)
ans =
      9     10
      2     20
      3     30

>> q(end-1,:)
ans =
      6      1      2     20
```

3.6   Deleting Rows or Columns

To get rid of a row or column set it equal to the empty matrix [].

```
>> a = [1 2 3;4 5 6;7 8 9]
a =
      1      2      3
      4      5      6
      7      8      9
>> a(:,2) = []
a =
      1      3
      4      6
      7      9
```

3.7   Matrix Arithmetic

Matrices can be added and subtracted (they must be the same size).

```
>> b = 10*a
b =
     10     30
     40     60
     70     90
```

```
>> a + b
ans  =
      11    33
      44    66
      77    99
```

3.8   Transpose

To convert rows into columns use the transpose symbol ':

```
>> a'
ans  =

       1     4     7
       3     6     9

>> b = [[1 2 3]' [4 5 6]']
b   =
       1     4
       2     5
       3     6
```

Be careful when taking the transpose of complex matrices. The transpose operator takes the complex conjugate transpose. If z is the matrix:

$$\begin{pmatrix} 1 & 0 - i \\ 0 + 2i & 1 + i \end{pmatrix}$$

then z' is:

$$\begin{pmatrix} 1 & 0 - 2i \\ 0 + i & 1 - i \end{pmatrix}.$$

To take the transpose without conjugating the complex elements, use the .' operator. In this case z.' is:

$$\begin{pmatrix} 1 & 0 + 2i \\ 0 - i & 1 + i \end{pmatrix}.$$

4   Basic Graphics

The bread-and-butter of MATLAB graphics is the plot command. Earlier we produced a plot of the sine function:

```
x = 0:.1:2*pi;
y = sin(x);
plot(x,y)
```

In this case we used `plot` to plot one vector against another. The elements of the vectors were plotted in order and joined by straight line segments. There are many options for changing the appearance of a plot. For example:

`plot(x,y,'r-.')`

will join the points using a red dash-dotted line. Other colours you can use are: `'c'`, `'m'`, `'y'`, `'r'`, `'g'`, `'b'`, `'w'`, `'k'`, which correspond to cyan, magenta, yellow, red, green, blue, white, and black. Possible line styles are: solid `'-'`, dashed `'--'`, dotted `':'`, and dash-dotted `'-.'`. To plot the points themselves with symbols you can use: dots `'.'`, circles `'o'`, plus signs `'+'`, crosses `'x'`, or stars `'*'`, and many others (type `help plot` for a list). For example:

`plot(x,y,'bx')`

plots the points using blue crosses without joining them with lines, and

`plot(x,y,'b:x')`

plots the points using blue crosses and joins them with a blue dotted line. Colours, symbols and lines can be combined, for example, `'r.-'`, `'rx-'` or `'rx:'`.

4.1   Plotting Many Lines

To plot more than one line you can specify more than one set of x and y vectors in the `plot` command:

`plot(x,y,x,2*y)`

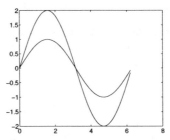

On the screen MATLAB distinguishes the lines by drawing them in different colours. If you need to print in black and white, you can differentiate the lines by plotting one of them with a dashed line:

`plot(x,y,x,2*y,'--')`

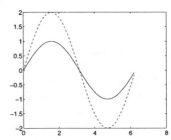

4.2   Adding Plots

When you issue a `plot` command MATLAB clears the axes and produces
a new plot. To add to an existing plot, type `hold on`. For example try
this:

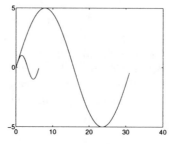

```
plot(x,y)
hold on
plot(5*x,5*y)
```

MATLAB re-scales the axes to fit the new data. The old plot appears
smaller. Once you have typed `hold on`, all subsequent plots will be
added to the current axes:

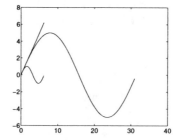

```
plot(x,x)
```

Companion M-Files Feature 1 *If you decide you want to re-
move the last thing you plotted on a plot with* hold *on in force,
you can type:*

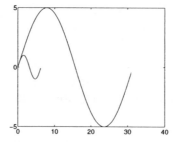

```
undo
```

to get back to where you were before.

To switch off the `hold` behaviour, type `hold off`. Typing `hold` by itself
toggles the hold state of the current plot.

4.3 Plotting Matrices

If one of the arguments to the plot command is a matrix, MATLAB will use the columns of the matrix to plot a set of lines, one line per column:

```
>> q = [1 1 1;2 3 4;3 5 7;4 7 10]
q =
       1    1    1
       2    3    4
       3    5    7
       4    7   10
>> plot(q)
>> grid
```

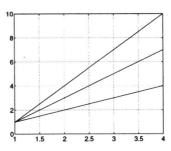

MATLAB plots the columns of the matrix q against the row index. You can also supply an x variable:

```
>> x = [0 1 3 6]
x =
       0    1    3    6
>> plot(x,q)
>> grid
```

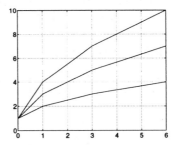

Here the x values are not uniformly spaced, but they are the same for each column of q. You can also plot a matrix of x values against a vector of y values (be careful: the y values are in the vector x):

```
plot(q,x)
grid
```

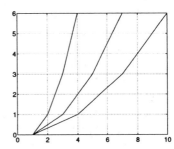

If both the x and y arguments are matrices, MATLAB will plot the successive columns on the same plot:

```
>> x = [[1 2 3 4]' [2 3 4 5]' [3 4 5 6]']
x  =
       1      2      3
       2      3      4
       3      4      5
       4      5      6
>> plot(x,q)
>> grid
```

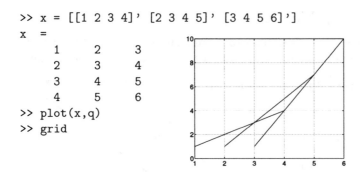

4.4   Clearing the Figure Window

You can clear the plot window by typing clf, which stands for 'clear figure'. To get rid of a figure window entirely, type close. To get rid of all the figure windows, type close all. New figure windows can be created by typing figure.

4.5   Subplots

To plot more than one set of axes in the same window, use the subplot command. You can type

```
subplot(m,n,p)
```

to break up the plotting window into m plots in the vertical direction and n plots in the horizontal direction, choosing the pth plot for drawing into. The subplots are counted as you read text: left to right along the top row, then left to right along the second row, and so on. Here is an example (do not forget to use the ↑ key to save typing):

```
t = 0:.1:2*pi;
subplot(2,2,1)
plot(cos(t),sin(t))
subplot(2,2,2)
plot(cos(t),sin(2*t))
subplot(2,2,3)
plot(cos(t),sin(3*t))
subplot(2,2,4)
plot(cos(t),sin(4*t))
```

If you want to clear one of the plots in a subplot without affecting the others you can use the cla (clear axes) command. Continuing the previous example:

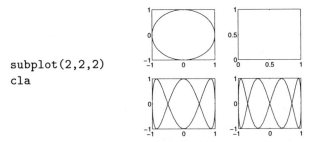

```
subplot(2,2,2)
cla
```

As long as your subplots are based on an array of 9 × 9 little plots or less, you can use a simplified syntax. For example, `subplot(221)` or `subplot 221` are equivalent to `subplot(2,2,1)`. You can mix different subplot arrays on the same figure, as long as the plots do not overlap:

```
subplot 221
plot(1:10)
subplot 222
plot(0,'*')
subplot 212
plot([1 0 1 0])
```

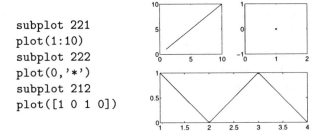

4.6   Three-Dimensional Plots

The `plot3` command is the 3-d equivalent of `plot`:

```
t = 0:.1:2*pi;
plot3(cos(3*t),sin(3*t),t)
```

The three dimensional spiral can be better visualised by changing the orientation of the axes. You can invoke a mouse-based 3-d axis mover by typing:

```
rotate3d
```

If you click the mouse button down on the plot and drag, you can move the axes and view the plot from any angle. Release the mouse button to redraw the data. Type `rotate3d` again to turn off this behaviour.

4.7   Axes

So far we have allowed MATLAB to choose the axes for our plots. You can change the axes in many ways:

axis([*xmin xmax ymin ymax*])	sets the axes' minimum and maximum values
axis square	makes the axes the same length
axis equal	makes the axes the same scale
axis tight	sets the axes limits to the range of the data
axis auto	allows MATLAB to choose axes limits
axis off	removes the axes leaving only the plotted data
axis on	puts the axes back again
grid on	draws dotted grid lines
grid off	removes grid lines
grid	toggles the grid
box*	toggles the box
zeroaxes*	draws the x-axis at $y = 0$ and vice-versa

The functions marked with an asterisk * are nonstandard features, implemented in this book's companion m-files.[1]

4.8   Labels

You can put labels, titles, and text on a plot by using the commands:

xlabel('*text*')	
ylabel('*text*')	
zlabel('*text*')	
title('*text*')	
text($x,y,$'*text*')	places text at position x,y
gtext('*text*')	use mouse to place text

To put mathematics in labels you can use MATLAB's backslash notation (familiar to users of the TEX typesetting system):

```
t = 0:.1:2*pi;
y1 = cos(t);
y2 = sin(t);
plot(t,y1,t,y2)
xlabel('0 \leq \theta < 2\pi')
ylabel('sin \theta, cos \theta')
text(1,cos(1),' cosine')
text(3,sin(3),' sine')
box
```

---

[1]MATLAB version 5.3 implements its own version of the box command.

Companion M-Files Feature 2 *To label many curves on a plot it is better to put the text close to the curves themselves rather than in a separate legend off to one side. Legends force the eye to make many jumps between the plot and the legend to sort out which line is which. Although* MATLAB *comes equipped with a* legend *function, I prefer to use the companion m-file* curlabel, *which is good especially for labelling plots which are close together:*

```
t = 0:.1:2*pi;
plot(t,sin(t),t,sin(1.05*t))
curlabel('frequency = 1')
curlabel('frequency = 1.05')
axis([0 max(t) -1 1])
zeroaxes
```

You must use the mouse to specify the start and end points of the pointer lines. The echo from the function can be pasted into an m-file for future use.

5 More Matrix Algebra

You can multiply two matrices together using the * operator:

```
>> a = [1 2;3 4]
a   =
        1       2
        3       4
>> b = [1 0 1 0;0 1 1 0]
b   =
        1       0       1       0
        0       1       1       0
>> a*b
ans   =
        1       2       3       0
        3       4       7       0

>> u = [1 2 0 1]
u   =
        1       2       0       1
>> v = [1 1 2 2]'
```

```
v   =
      1
      1
      2
      2
>> v*u
ans  =
      1      2      0      1
      1      2      0      1
      2      4      0      2
      2      4      0      2
>> u*v
ans  =
      5
```

The matrix inverse can be found with the `inv` command:

```
>> a = pascal(3)
a   =
      1      1      1
      1      2      3
      1      3      6
>> inv(a)
ans  =
      3     -3      1
     -3      5     -2
      1     -2      1
>> a*inv(a)
ans  =
      1      0      0
      0      1      0
      0      0      1
```

To multiply the elements of two matrices use the `.*` operator:

```
>> a = [1 2;3 4]
a   =
      1      2
      3      4
>> b = [2 3;0 1]
b   =
      2      3
      0      1

>> a.*b
ans  =
      2      6
      0      4
```

To raise the elements of a matrix to a power use the `.^` operator:

```
>> a = pascal(3)
a   =
        1       1       1
        1       2       3
        1       3       6
>> a.^2
ans   =
        1       1       1
        1       4       9
        1       9      36
```

6   Basic Data Analysis

The following functions can be used to perform data analysis functions:

max	maximum
min	minimum
find	find indices of nonzero elements
mean	average or mean
median	median
std	standard deviation
sort	sort in ascending order
sortrows	sort rows in ascending order
sum	sum of elements
prod	product of elements
diff	difference between elements
trapz	trapezoidal integration
cumsum	cumulative sum
cumprod	cumulative product
cumtrapz	cumulative trapezoidal integration

As we have seen with the `plot` command, MATLAB usually prefers to work with matrix columns, rather than rows. This is true for many of MATLAB's functions, which work on columns when given matrix arguments. For example:

```
>> a = magic(3)
a   =
        8       1       6
        3       5       7
        4       9       2
>> m = max(a)
m   =
        8       9       7
```

`max` returns a vector containing the maximum value of each column. When given a vector, `max` returns the maximum value:

```
>> max(m)
ans  =
     9
```

To find the index corresponding to the maximum value, supply two output arguments to `max`:

```
>> [v,ind] = max(m)
v  =
     9
ind  =
     2
```

The first argument is the maximum value and the second is the index of the maximum value. Another example is

```
>> x = 0:.01:2;
>> y = humps(x);
>> plot(x,y)
>> [v,ind] = max(y)
v  =
    96.5000
ind  =
    31
>> hold on
>> plot(x(ind),y(ind),'ro')
>> x(ind)
ans  =
     0.3000
>> y(ind)
ans  =
    96.5000
```

The `find` function is often used with relational and logical operators:

Relational operators		
	==	equal to
	~=	not equal to
	<	less than
	>	greater than
	<=	less than or equal to
	>=	greater than or equal to

Logical operators	&	AND
	|	OR
	~	NOT
	xor	EXCLUSIVE OR
	any	True if any element is non-zero
	all	True if all elements are non-zero

We continue the previous example and use `find` to plot the part of the peaks function that lies between $y = 20$ and $y = 40$:

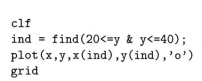

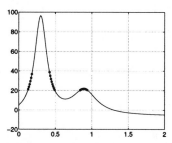

When used with one output argument, `find` assumes that the input is a vector. When the input is a matrix `find` first strings out the elements as a single column vector and returns the corresponding indices. As an example we consider the spiral matrix:

```
>> s = spiral(3)
s =
     7     8     9
     6     1     2
     5     4     3
```

We find the elements of s less than 6:

```
>> s<6
ans =
     0     0     0
     0     1     1
     1     1     1
>> find(s<6)
ans =
     3
     5
     6
     8
     9
```

The result of `find` is a vector of indices of s counted down the first column, then the second, and then the third. The following example shows how the results of the find command can be used to extract elements from a matrix that satisfy a logical test:

```
>> s = 100*spiral(3)
s  =
    700     800     900
    600     100     200
    500     400     300
>> ind = find(s>400)
ind  =
      1
      2
      3
      4
      7
>> s(ind)
ans  =
    700
    600
    500
    800
    900
>> s(s>400)
ans  =
    700
    600
    500
    800
    900
```

After introducing graphics of functions of two variables in the next section, we will see how the find command can be used to do the three-dimensional equivalent of the plot shown on page 23, where the domain of a curve satisfying a logical test was extracted.

7   Graphics of Functions of Two Variables

7.1   Basic Plots

A MATLAB surface is defined by the z coordinates associated with a set of (x, y) coordinates. For example, suppose we have the set of (x, y) coordinates:

$$(x, y) = \begin{pmatrix} 1,1 & 2,1 & 3,1 & 4,1 \\ 1,2 & 2,2 & 3,2 & 4,2 \\ 1,3 & 2,3 & 3,3 & 4,3 \\ 1,4 & 2,4 & 3,4 & 4,4 \end{pmatrix}.$$

The points can be plotted as (x, y) pairs:

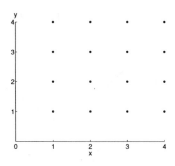

The (x, y) pairs can be split into two matrices:

$$x = \begin{pmatrix} 1 & 2 & 3 & 4 \\ 1 & 2 & 3 & 4 \\ 1 & 2 & 3 & 4 \\ 1 & 2 & 3 & 4 \end{pmatrix}; \quad y = \begin{pmatrix} 1 & 1 & 1 & 1 \\ 2 & 2 & 2 & 2 \\ 3 & 3 & 3 & 3 \\ 4 & 4 & 4 & 4 \end{pmatrix}.$$

The matrix x varies along its columns and y varies down its rows. We define the surface z:

$$z = \sqrt{x^2 + y^2};$$

which is the distance of each (x, y) point from the origin $(0, 0)$. To calculate z in MATLAB for the x and y matrices given above, we begin by using the meshgrid function, which generates the required x and y matrices:

```
>> [x,y] = meshgrid(1:4)
x   =
        1       2       3       4
        1       2       3       4
        1       2       3       4
        1       2       3       4
y   =
        1       1       1       1
        2       2       2       2
        3       3       3       3
        4       4       4       4
```

Now we simply convert our distance equation to MATLAB notation; $z = \sqrt{x^2 + y^2}$ becomes:

```
>> z = sqrt(x.^2 + y.^2)
z   =
        1.4142      2.2361      3.1623      4.1231
        2.2361      2.8284      3.6056      4.4721
        3.1623      3.6056      4.2426      5.0000
        4.1231      4.4721      5.0000      5.6569
```

We can plot the surface z as a function of x and y:

mesh(x,y,z)

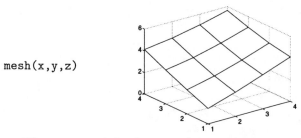

We can expand the domain of the calculation by increasing the input to meshgrid. Be careful to end the lines with a semicolon to avoid being swamped with numbers:

```
[x,y] = meshgrid(-10:10);
z = sqrt(x.^2 + y.^2);
mesh(x,y,z)
```

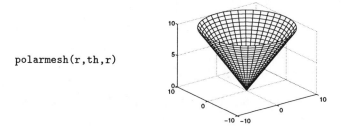

The surface is an inverted cone, with its apex at $(0,0,0)$.

Companion M-Files Feature 3 *A clearer plot can be produced using a polar grid, instead of a rectilinear grid. We can use the companion function* **polarmesh** *to produce such a plot. First we define a polar grid of points:*

```
[r,th] = meshgrid(0:.5:10,0:pi/20:2*pi);
```

Then display the surface defined by $z = r$:

polarmesh(r,th,r)

A more interesting surface is

$$z = 3(1-x)^2 e^{-x^2-(y+1)^2} - 10(\tfrac{1}{5}x - x^3 - y^5)e^{-x^2-y^2} \cdots$$
$$- \tfrac{1}{3}e^{-(x+1)^2-y^2} .$$

In MATLAB notation you could type:

```
z =   3*(1-x).^2.*exp(-(x.^2) - (y+1).^2) ...
    - 10*(x/5 - x.^3 - y.^5).*exp(-x.^2-y.^2) ...
    - 1/3*exp(-(x+1).^2 - y.^2);
```

but you do not have to type this because it is already defined by the function peaks. Before plotting we define the data and set the colour map to gray:

```
[x,y,z] = peaks;
colormap(gray)
```

The following plots show 10 different ways to view this data.

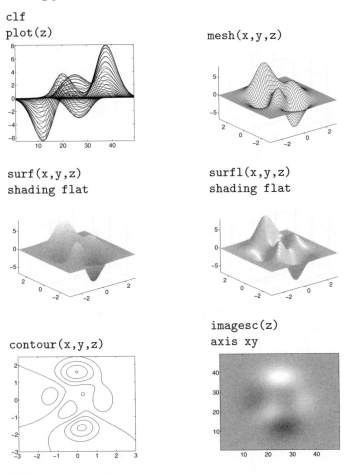

surfc(x,y,z)

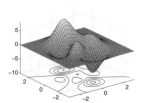

contourf(x,y,z)

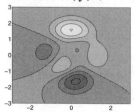

plot3(x,y,z,'k')
hold on
contour3(x,y,z,'k')

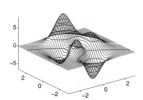

clf
spanplot(z)

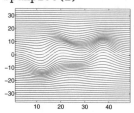

The contour function plots the contours using the current colour map's colours (see next section). Adding the specifier 'k' to the end of the argument list draws the contours in black. The spanplot function is nonstandard and is included in the companion software.

You should experiment with these plots. Try typing help for each of these plot commands. Explore the various ways of shading a surface, try using different colour maps (see next section) or viewing angles (help view), or try modifying the surface and replotting. Remember that rotate3d can be used to switch on a click-and-drag three-dimensional view changer: click down on the plot and drag it to alter the viewing angle; release the mouse to redraw the plot. (If rotate3d is already switched on, typing rotate3d again will switch it off.)

7.2   Colour Maps

MATLAB uses a matrix called a colour map to apply colour to surfaces and images. The idea is that different colours will be used to draw various parts of the plot depending on the colour map. The colour map is a list of triplets corresponding to the intensities of the red, green, and blue video components, which add up to yield other colours. The intensities must be between zero and one. Some example colours are shown in this table.

Red	Green	Blue	Colour·
0	0	0	Black
1	1	1	White
1	0	0	Red
0	1	0	Green
0	0	1	Blue
1	1	0	Yellow
1	0	1	Magenta
0	1	1	Cyan
.5	.5	.5	Gray
.5	0	0	Dark red
1	.62	.4	Dark orange
.49	1	.83	Aquamarine
.95	.9	.8	Parchment

Yellow, for example, consists of the combination of the full intensities of red and green, with no blue, while gray is the combination of 50% intensities of red, green, and blue.

You can create your own colour maps or use any of MATLAB's many predefined colour maps:

```
hsv      hot      gray     bone       copper   pink
white    flag     lines    colorcube  jet      prism
cool     autumn   spring   winter     summer
```

Two nonstandard colour maps that are supplied in the companion software include `redblue` and `myjet`. The first consists of red blending to blue through shades of gray. The second consists of a modification of the `jet` colour map that has white at the top instead of dark red.

These functions all take an optional parameter that specifies the number of rows (colours) in the colour map matrix. For example, typing `gray(8)` creates an 8×3 matrix of various levels of gray:

```
>> gray(8)
ans  =
         0        0        0
    0.1429   0.1429   0.1429
    0.2857   0.2857   0.2857
    0.4286   0.4286   0.4286
    0.5714   0.5714   0.5714
    0.7143   0.7143   0.7143
    0.8571   0.8571   0.8571
    1.0000   1.0000   1.0000
```

To tell MATLAB to use a colour map, type it as an input to the `colormap` function:

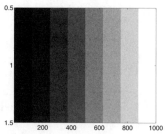

```
m = gray(8);
colormap(m)
imagesc(1:1000)
```

Most of MATLAB's surface viewing functions use the colour map to apply colour to the surface depending on the z-value. The `imagesc` function produces a coloured image of the matrix argument, colouring each element depending on its value. The smallest element will take the colour specified in the first row of the colour map, the largest element will take the colour specified in the last row of the colour map, and all the elements in between will take linearly interpolated colours.

To get a plot of the levels of red, green, and blue in the current colour map use `rgbplot`:

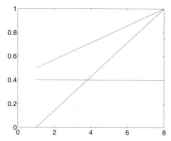

```
colormap(summer)
rgbplot(colormap)
```

On the screen the lines corresponding to the red, green, and blue components of the colour map are coloured red, green, and blue, respectively.

7.3   Colour Bar

To display the current colour map use the `colorbar` function:

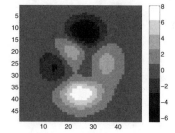

```
z = peaks;
colormap(gray(8))
imagesc(z)
colorbar
```

7.4   Good and Bad Colour Maps

Much research has been done on human perception of colours and, in particular, how different viewers interpret coloured images as value-scales.

The conclusion is that most viewers find it very difficult to interpret these sorts of images; the cognitive switch from, for example, ROYG-BIV to amplitude is very slow and nonintuitive. A way out of this is to use a palette of slightly varying, nonsaturated colours. These sorts of colours have been used to create high-quality geographic maps for many years. Most of MATLAB's colour maps consist of highly saturated colours (including the default colour map, which is jet(64)). It is better to forgo these sorts of colour maps and stick with the calmer ones such as gray, bone, or summer. The gray colour map has the added advantage that printed versions will reproduce easily, for example, on a photocopier.[2] The companion m-files include some other colour maps: redblue, myjet, yellow, green, red, and blue.

To distinguish adjacent patches of subtly different colours, the eye can be helped by enclosing the patches with a thin dark edge. The contourf function, therefore, is an excellent way of displaying this sort of data.[3]

7.5 Extracting Logical Domains

Let us look again at the peaks function:

```
[x,y,z] = peaks;
surfl(x,y,z)
axis tight
colormap(gray(64))
```

Suppose we want to extract the part of this surface for which the z values lie between 2 and 4. We use exactly the same technique as was given on page 23. The find command is used first to find the indices of the z values that satisfy the logical test:

```
>> ind = find(2<=z & z<=4);
>> size(ind)
ans =
    234    1
```

There are 234 elements in z that satisfy our condition. We can plot these elements over the surface as follows:

---

[2]Edward R. Tufte, *Visual Explanations* (Graphics Press, Cheshire Connecticut, 1997), pp. 76–77.

[3]Edward R. Tufte, *Envisioning Information* (Graphics Press, Cheshire Connecticut, 1990), pp. 88ff.

```
hold on
plot3(x(ind),y(ind),z(ind),'.')
```

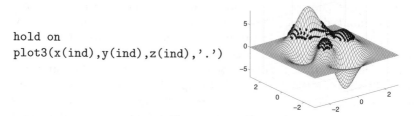

The x, y domain of the extracted points can be shown clearly with an overhead view:

```
view(2)
xyz
shading flat
```

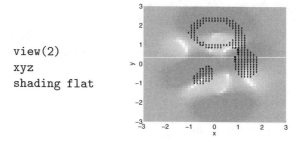

The associated z values can be shown with a side view:

```
view(90,0)
grid
```

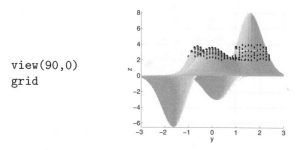

7.6   Nonrectangular Surface Domains

The `polarmesh` function given on page 26 showed a conical function defined over a circular domain of x and y points. Let us now look a bit more generally at how to define such nonrectangular domains for surfaces.

The standard MATLAB functions, including graphics functions, tend to like working with rectangular matrices: each row must have the same number of columns. For surfaces, this requirement applies to the x, y and z matrices that specify the surface. Let us demonstrate by way of an example. First we generate a rectangular domain of x and y points, with x going from -1 to 1, and y going from 0 to 2:

```
>> [x,y] = meshgrid(-1:1,1:3)
```

```
x   =
        -1       0       1
        -1       0       1
        -1       0       1
y   =
         1       1       1
         2       2       2
         3       3       3
```

This set of points defines a rectangular domain because the rows of x are identical and the columns of y are identical. We can make a plot of the points (as we did on page 25):

```
clf
plot(x,y,'.')
```

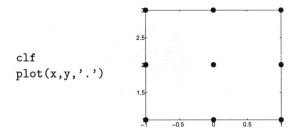

Now let us change the y matrix a bit:

```
>> y = [[1; 2; 3] [1; 1.5; 2] [0; .2; .4]]
y   =
        1.0000        1.0000             0
        2.0000        1.5000        0.2000
        3.0000        2.0000        0.4000
```

The plot of this data looks like a bent triangle:

```
plot(x,y,'.')
```

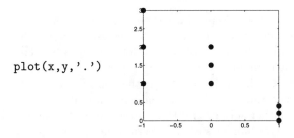

To define a surface over this domain we simply have to supply the z values. We can start by simply defining a constant z:

```
>> z = 5*ones(3,3)
z  =

       5      5      5
       5      5      5
       5      5      5
>> surf(x,y,z)
```

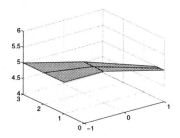

But, of course, the z values need not be constant:

```
>> z = [[4; 5; 6] [4.5; 5; 5.5] [4.9; 5; 5.1]]
z  =
      4.0     4.5     4.9
      5.0     5.0     5.0
      6.0     5.5     5.1
>> surf(x,y,z)
```

Other graphics functions can also handle nonrectangular grids. Here is an example using the `contour` function:

```
cs = contour(x,y,z,'k');
clabel(cs)
i = [1 4 7 9 6 3 1];
hold on
plt(x(i),y(i),':')
```

The contour levels are labelled using the `clabel` command, and the region defined by the x and y points is outlined by the dotted line. The contours that the labels refer to are marked by small plus signs '+'. The outline around the bent domain is drawn using the x and y matrices indexed using the vector i. The vector i extracts the appropriate points from the x and y matrices using the columnar indexing described in section 3.4 on page 9. The other surface graphics functions—mesh, `surfl`, `surfc`, and `contourf`—can handle such nonrectangular grids equally well. The `image` and `imagesc` functions assume equally spaced rectangular grids and cannot handle anything else. (The `pcolor` function draws a surface and sets the view point to directly overhead, so it is not discussed separately.)

Let us now do another example of a surface defined over a non-rectangular grid. We want to define a set of points that cover the semi-

annular region as shown in the diagram
at right. To define such a set of points
we use a polar grid based on radial and
angular coordinates r and θ. We use the
following limits on these coordinates:

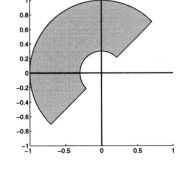

$$.3 \leq r \leq 1$$
$$\pi/4 \leq \theta \leq 5\pi/4$$

These are set up in MATLAB as follows:

```
rv = linspace(.3,1,50);
thv = linspace(pi/4,5*pi/4,50);
[r,th] = meshgrid(rv,thv);
```

where the calls to linspace produce vectors of 50 points covering the
intervals. The x and y points are defined by the following trigonometric
relations:

```
x = r.*cos(th);
y = r.*sin(th);
```

Now our semi-annular region is defined. To prove it, let us plot the
points:

```
plot(x,y,'.')
```

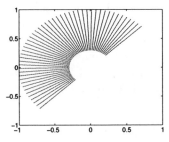

Again, we can define any z matrix we like. Just for fun, we use the peaks
function of the right size and add a linear ramp:

```
z = peaks(50) + 10*x;
surf(x,y,z)
```

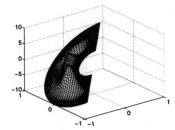

As we did in the previous example, we check that the contour function
works (omitting the labels this time, and upping the number of contours
drawn to 30):

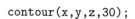 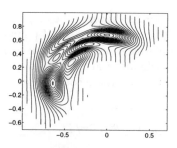

`contour(x,y,z,30);`

You may have noticed that the semi-annular region does not appear as a circular segment in our plots. That is because the axes are not square. To get square axes you can use the `axis square` command as described on pages 18 and 120.

In this section we have looked at surfaces having domains that could be defined in terms of rectangular x and y data matrices. Domains that cannot be defined with such matrics are discussed in section 36 on page 157. For example all x values may not have the same number of y values, or the x, y points could be scattered about in an irregular way.

8   M-Files

Until now we have driven MATLAB by typing in commands directly. This is fine for simple tasks, but for more complex ones we can store the typed input into a file and tell MATLAB to get its input from the file. Such files must have the extension ".m". They are called m-files. If an m-file contains MATLAB statements just as you would type them into MATLAB, they are called *scripts*. M-files can also accept input and produce output, in which case they are called *functions*.

8.1   Scripts

Using your text editor create a file called `mfile1.m` containing the following lines:

```
z = peaks;
zplot = z;

% Do the peaks:

clf
subplot(221)
ind = find(z<0);
zplot(ind) = zeros(size(ind));
mesh(zplot)
axis tight
```

```
% Do the valleys:

subplot(222)
ind = find(z>0);
zplot = z;
zplot(ind) = zeros(size(ind));
mesh(zplot)
axis tight
```

Now try this in the MATLAB window:

mfile1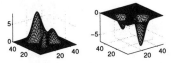

MATLAB has executed the instructions in `mfile1.m` just as if you had typed them in. The lines beginning with the percent sign % are ignored by MATLAB so they can be used to put comments in your code. Blank lines can be used to improve readability.

Any variables created by a script m-file are available in the command window after the m-file completes running:

```
>> clear
>> whos
>> mfile1
>> whos
  Name       Size        Bytes  Class
  ind        1544x1      12352  double array
  z          49x49       19208  double array
  zplot      49x49       19208  double array

Grand total is 6346 elements using 50768 bytes
```

These variables are said to exist in the MATLAB *workspace*. Scripts can also operate on variables that already exist in the workspace.

You can type the name of a script file within another script file. For example you could create another file called `mfile2` that contains the text line `mfile1`; the contents of `mfile1` will then be executed at that point within `mfile2`.

8.2 Functions

Functions are m-files that can be used to extend the MATLAB language. Functions can accept input arguments and produce output arguments. Many of MATLAB's own commands are implemented as m-files; try typing `type mean` to see how MATLAB calculates the mean. Functions use

variables that are local to themselves and do not appear in the main
workspace. This is an example of a function:

```
function x = quadratic(a,b,c)

% QUADRATIC Find roots of a quadratic equation.
%
% X = QUADRATIC(A,B,C) returns the two roots of the
% quadratic equation
%
%                   y = A*x^2 + B*x + C.
%
% The roots are contained in X = [X1 X2].

% A. Knight, July 1997
delta = 4*a*c;
denom = 2*a;
rootdisc = sqrt(b.^2 - delta); % Root of the discriminant
x1 = (-b + rootdisc)./denom;
x2 = (-b - rootdisc)./denom;
x = [x1 x2];
```

Function m-files must start with the word `function`, followed by
the output variable(s), an equals sign, the name of the function, and
the input variable(s). Functions do not have to have input or output
arguments. If there is more than one input or output argument, they
must be separated by commas. If there are one or more input arguments,
they must be enclosed in brackets, and if there are two or more output
arguments, they must be enclosed in square brackets. The following
illustrate these points (they are all valid function definition lines):

```
function [xx,yy,zz] = sphere(n)
function fancyplot
function a = lists(x,y,z,t)
```

Function names must follow the same rules as variable names. The file
name is the function name with ".m" appended. If the file name and
the function name are different, MATLAB uses the file name and ignores
the function name. You should use the same name for both the function
and the file to avoid confusion.

Following the function definition line you should put comment lines
that explain how to use the function. These comment lines are printed in
the command window when you type `help` followed by the m-file name
at the prompt:

```
>> help quadratic
  QUADRATIC Find roots of a quadratic equation.
```

```
X = QUADRATIC(A,B,C) returns the two roots of the
quadratic equation
                 y = A*x^2 + B*x + C.
The roots are contained in X = [X1 X2].
```

MATLAB only echoes the comment lines that are contiguous; the first non-comment line, in this case the blank line before the signature, tells MATLAB that the help comments have ended. The first line of the help comments is searched and, if successful, displayed when you type a `lookfor` command.

Comment lines can appear anywhere in the body of an m-file. Comments can be put at the end of a line of code:

```
rootdisc = sqrt(b.^2 - delta); % Root of the discriminant
```

Blank lines can appear anywhere in the body of an m-file. Apart from ending the help comment lines in a function, blank lines are ignored.

8.3   Flow Control

MATLAB has four kinds of statements you can use to control the flow through your code:

`if, else` and `elseif`   execute statements based on a logical test
`switch, case` and `otherwise`   execute groups of statements based on a logical test
`while` and `end`   execute statements an indefinite number of times, based on a logical test
`for` and `end`   execute statements a fixed number of times

If, Else, Elseif

The basic form of an `if` statement is:

```
    if test
        statements
    end
```

The *test* is an expression that is either 1 (true) or 0 (false). The *statements* between the `if` and `end` statements are executed if the *test* is true. If the *test* is false the *statements* will be ignored and execution will resume at the line after the `end` statement. The *test* expression can be a vector or matrix, in which case *all* the elements must be equal to 1 for the *statements* to be executed. Further tests can be made using the `elseif` and `else` statements.

> **Exercise 2** *Write a function m-file that takes a vector input and returns 1 if all of the elements are positive, −1 if they are all negative, and zero for all other cases. Hint: Type* **help all**. *(Answer on page 183.)*

Switch

The basic form of a switch statement is:

```
switch test
      case result1
            statements
      case result2
            statements

            .

            .

            .


      otherwise
            statements
   end
```

The respective *statements* are executed if the value of *test* is equal
to the respective *results*. If none of the cases are true, the otherwise
statements are done. Only the first matching case is carried out. If
you want the same *statements* to be done for different cases, you can
enclose the several *results* in curly brackets:

```
switch x
  case 1
    disp('x is 1')
  case {2,3,4}
    disp('x is 2, 3 or 4')
  case 5
    disp('x is 5')
  otherwise
    disp('x is not 1, 2, 3, 4 or 5')
end
```

While

The basic form of a while loop is

```
while test
        statements
   end
```

The *statements* are executed repeatedly while the value of *test* is
equal to 1. For example, to find the first integer n for which $1+2+\cdots+n$
is is greater than 1000:

```
n = 1;
while sum(1:n)<=1000
  n = n+1;
end
```

A quick way to 'comment out' a slab of code in an m-file is to enclose it between a `while` 0 and `end` statements. The enclosed code will never be executed.

For

The basic form of a `for` loop is:

```
for index = start:increment:stop
    statements
end
```

You can omit the increment, in which case an increment of 1 is assumed. The increment can be positive or negative. During the first pass through the loop the *index* will have the value *start*. The *index* will be increased by *increment* during each successive pass until the *index* exceeds the value *stop*. The following example produces views of the peaks function from many angles:

```
clf
colormap(gray)
plotnum = 1;
z = peaks(20);
for az = 0:10:350
   subplot(6,6,plotnum)
   surfl(z),shading flat
   view(az,30)
   axis tight
   axis off
   plotnum = plotnum + 1;
end
```

The index of a `for` loop can be a vector or a matrix. If it is a vector the loop will be done as many times as the number of elements in the vector, with the index taking successive values of the vector in each pass. If the index is a matrix, the loop will be done as many times as there are columns in the matrix, with the index taking successive columns of the matrix in each pass. For example:

```
>> q = pascal(3)
q =
      1     1     1
      1     2     3
      1     3     6
>> for i = q,i,end
i =
      1
      1
      1
```

```
i  =
      1
      2
      3
i  =
      1
      3
      6
```

Vectorised Code

MATLAB is a matrix language, and many of its algorithms are optimised
for matrices. MATLAB code can often be accelerated by replacing `for`
and `while` loops with operations on matrices. In the following example,
we calculate the factorial of the numbers from 1 to 500 using a `for`
loop. Create a script m-file called `factorialloop.m` that contains the
following code:

```
for number = 1:500
  fact = 1;
  for i = 2:number
    fact = fact*i;
  end
  y(number) = fact;
end
```

We can time how long this program takes to run by using the stopwatch
functions `tic` and `toc`:

```
>> tic;factorialloop;toc
elapsed_time =
    4.6332
```

which is the time in seconds. The same calculation can be done in much
less time by replacing the internal `for` loop by the `prod` function. Create
an m-file called `factorialvect.m`:

```
for number = 1:500
  y(number) = prod(1:number);
end
```

This version takes about a tenth of the time:

```
>> clear
>> tic;factorialvect;toc
elapsed_time =
    0.4331
```

Further increases in speed can be achieved by pre-allocating the output matrix y. If we have an m-file called `factorialpre.m`:

```
y = zeros(1,500);
for number = 1:500
  y(number) = prod(1:number);
end
```

the execution time is about 10% faster:[4]

```
>> clear
>> tic;factorialpre;toc
elapsed_time  =
    0.3752
```

More on vectorising code is given in Part II (see page 175).

8.4   Comparing Strings

The tests in flow control statements often involve strings (arrays of characters). For example you may want to ask the user of an m-file a question which has a "yes" or "no" response, and adjust the flow accordingly. Although MATLAB has sophisticated menu utilities, the following is often sufficient to get a user input:

```
input('Do you want to continue (y or n) ? ','s');
```

The `'s'` at the end tells MATLAB to expect a string response, rather than a numerical response. The following MATLAB code tests for a 'y' response:

```
if strcmp(lower(ans(1)),'y')
    go_ahead
else
    return
end
```

The `strcmp` function compares strings, `lower` converts to lower-case characters and `ans(1)` selects the first letter of the response. Type `help strcmp` for more information. The `return` command returns to the invoking function or to the MATLAB prompt.

9   Data Files

Many techniques are available to read data into MATLAB and to save data from MATLAB. The `load` and `save` functions can load or save MATLAB format binary or plain ASCII files, and low-level input-output routines can be used for other formats.

---

[4]See MATLAB's **gamma** function if you are interested in computing factorials.

9.1   MATLAB Format

To save all the variables in the workspace onto disk use the `save` command. Typing `save keepfile` will save the workspace variables to a disk file called `keepfile.mat`, a binary file whose format is described in the MATLAB documentation. This data can be loaded into MATLAB by typing `load keepfile`.

To save or load only certain variables, specify them after the filename. For example, `load keepfile x` will load only the variable x from the saved file. The wild-card character * can be used to save or load variables according to a pattern. For example, `load keepfile *_test` loads only the variables that end with `_test`.

When the filename or the variable names are stored in strings, you can use the functional forms of these commands, for example:

```
save keepfile      is the same as   save('keepfile')
save keepfile x    ...              save('keepfile','x')
load keepfile      ...              A = 'keepfile'
                                    load(A)
```

Exercise 3 *The file* `clown.mat` *contains an image of a clown. What colour is his hair? (Answer on page 183.)*

9.2   ASCII Format

A file containing a list or table of numbers in ASCII format can be loaded into MATLAB. The variable containing the data is given the same name as the file name without the extension. For example, if a file `nums.dat` contained ASCII data, `load nums.dat` would load the data into a variable called `nums`. If the ASCII file contained a table of numbers, the variable would be a matrix the same size as the table.

Other functions are available to read various forms of delimiter-separated text files:

```
csvread    Read a comma separated value file
csvwrite   Write a comma separated value file
dlmread    Read ASCII delimited file
dlmwrite   Write ASCII delimited file
```

9.3   Other Formats

MATLAB's low-level input/output routines can be used to access more unusual data formats. They are listed here for reference:

File Opening and Closing:	`fclose`	`fopen`
Unformatted I/O:	`fread`	`fwrite`
Formatted I/O:	`fgetl`	`fprintf`
	`fgets`	`fscanf`
File Positioning:	`feof`	`fseek`
	`ferror`	`ftell`
	`frewind`	
String Conversion:	`sprintf`	`sscanf`

10 Directories

When you type a string of characters, say `asdf` at the MATLAB prompt and press return, MATLAB goes through the following sequence to try to make sense of what you typed:

1. Look for a variable called `asdf`;

2. Look for a built in MATLAB function called `asdf`;

3. Look in the current directory for an m-file called `asdf.m`;

4. Look in the directories specified by the MATLAB search path for an m-file called `asdf.m`.

The following commands are useful for working with different directories in MATLAB:

`cd`	Change to another directory
`pwd`	Display (print) current working directory
`dir`	Display contents of current working directory
`what`	Display MATLAB-relevant files
	in current working directory
`which`	Display directory containing specified function
`type`	Display file in the MATLAB window
`path`	Display or change the search path
`addpath`	Add directory to the search path
`rmpath`	Remove directory from the search path

If the directory name contains a blank space, enclose it in single quotes:

```
dir 'my documents'
```

(On PCs or Macintoshes you can use the Path Browser GUI to manipulate the path. Select 'File'→'Set Path' or click the Path Browser button on the tool bar.)

11   Startup

Each time you start MATLAB it looks for a script m-file called `startup.m`
and, if it finds it, does it. Thus, you can use `startup.m` to do things like
set the search path, set command and figure window preferences (e.g.,
set all your figures to have a black background), etc.

On PCs you should put the `startup.m` file in the directory called
`C:\MATLAB\toolbox\local`. On UNIX workstations you should put
your startup file in a directory called `matlab` immediately below your
home directory: `~/matlab`.

12   Using MATLAB on Different Platforms

A MATLAB format binary (`.mat`) file that is saved on one platform (say,
a PC or a Macintosh) can be transferred to a different platform (say, a
Unix or VMS box) and loaded into MATLAB running on that platform.
The mat-file contains information about the platform that saved the
data. MATLAB checks to see if the file was saved on a different platform,
and performs any necessary conversions automatically.

MATLAB m-files are ordinary ASCII text, and are immediately trans-
portable between platforms. Different platforms may use different characters to terminate lines of text (with CR and LF characters), but MAT-
LAB handles them all. However, the text editor you use must be able to
handle the end-of-line characters correctly.

The program you use to transfer m-files or mat-files, for example,
FTP or mail, must do so without corrupting the data. For FTP, for
example, mat-files must be transmitted in *binary* mode and m-files must
be transmitted in *ASCII* mode.

13   Log Scales

When dealing with data that varies over several orders of magnitude
a plain linear plot sometimes fails to display the variation in the data.
For example, consider the census estimates[5] of Australia's European
population at various times. If this data is contained in the file
`population.dat`, we can load and plot it as follows:

---

[5]Australian Bureau of Statistics Web Page, `http://www.statistics.gov.au`,
and *Australians: A Historical Library, Australians: Historical Statistics*, Fairfax,
Syme & Weldon Associates, 235 Jones Street, Broadway, New South Wales 2007,
Australia, 1987, pp. 25,26.

```
load population.dat
year = population(:,1);
P = population(:,2);
plot(year,P,':o')
box;grid
```

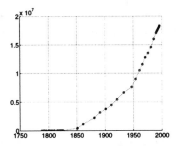

The European population prior to 1850 was very low and we are unable to see the fine detail. Detail is revealed when we use a logarithmic y-scale:

```
semilogy(year,P,':o')
box;grid
```

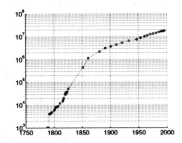

The following functions implement logarithmic axes:

loglog	Both axes logarithmic
semilogx	logarithmic x-axis
semilogy	logarithmic y-axis

14 Curve Fitting—Matrix Division

We continue with the example of Australian population data given in the previous section. Let us see how well a polynomial fits this data. We assume the data can be modelled by a parabola:

$$p = c_0 + c_1 x + c_2 x^2$$

where x is the year, c_0, c_1, and c_2 are coefficients to be found, and p is the population. We write down this equation substituting our measured data:

$$p_1 = c_0 + c_1 x_1 + c_2 x_1^2$$
$$p_2 = c_0 + c_1 x_2 + c_2 x_2^2$$
$$\vdots$$
$$p_N = c_0 + c_1 x_N + c_2 x_N^2$$

Where p_i is the population for year x_i, and $i = 1, 2, \ldots N$. We can write this series of equations as a matrix equation:

$$
\begin{pmatrix} p_1 \\ p_2 \\ \vdots \\ p_N \end{pmatrix} = \begin{pmatrix} 1 & x_1 & x_1^2 \\ 1 & x_2 & x_2^2 \\ & \vdots & \\ 1 & x_N & x_N^2 \end{pmatrix} \begin{pmatrix} c_0 \\ c_1 \\ c_2 \end{pmatrix}.
$$

Or, defining matrices,

$$
\mathbf{P} = \mathbf{X} \cdot \mathbf{C}.
$$

In MATLAB the $\mathbf{X}$ matrix is calculated as follows:

```
>> X = [ones(size(year)) year year.^2]
X  =
            1         1788      3196944
            1         1790      3204100
            .
            .
            .
            1         1993      3972049
            1         1994      3976036
            1         1995      3980025
```

The backslash operator solves the equation for the coefficient matrix $\mathbf{C}$:

```
>> C = X\P
C  =
    1.0e+09 *
    2.0067
   -0.0022
    0.0000
```

The third coefficient is not really zero; it is simply too small (compared to 2.0×10^9) to show in the default output format. We can change this by typing:

```
>> format long e
>> C
C  =
    2.006702229622023e+09
   -2.201930087288049e+06
    6.039665477603122e+02
```

The backslash operator does its best to solve a system of linear equations using Gaussian elimination or least-squares algorithms, depending on whether the system is exact, or over- or under-determined. We can display the resulting fit to the data by calculating the parabola. We use matrix multiplication to calculate the polynomial over a fine set of points separated by half a year:

```
year_fine = (year(1):0.5:year(length(year)))';
Pfine = [ones(size(year_fine)) year_fine year_fine.^2]*C;
```

```
plot(year,P,'o',...
     year_fine,Pfine)
```

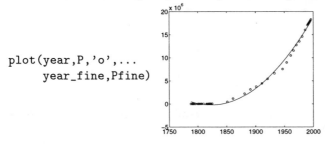

This technique can be used to fit any function that is linear in its parameters. (MATLAB provides the functions polyfit and polyval as easy interfaces to the functionality that we have just illustrated using matrix multiplication and division.)

> **Exercise 4** *Use this technique to fit an exponential curve to the population data. Hint: Take logs. (Answer on page 183.)*

15 Missing Data

Real-world measurements are often taken at regular intervals; for example, the position of a comet in the sky measured each night, or the depth of the sea along a line at 1 metre increments. Environmental effects or equipment failure (a cloudy night or a failed depth meter) sometimes result in a set of data that has missing values. In MATLAB these can be represented by NaN, which stands for "not-a-number". NaN is also given by MATLAB as the result of undefined calculations such as $0/0$. MATLAB handles NaNs by setting the result of any calculation that involves NaNs to NaN. Let us look at an example:

```
y = [1:4 NaN 6:14 NaN 16:20];
plot(y,'o')
grid;box
```

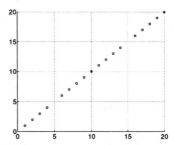

In everyday language we would say that the fifth and the fifteenth values of the y-vector are missing. MATLAB's graphics functions usually handle

NaNs by leaving them off the plot. For example, if we allow `plot` to try to join the points with a straight line, the values on either side of the NaNs terminate the line:

`plot(y)`
`grid;box`

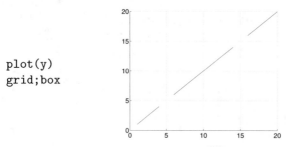

If we calculate the difference between y-values, the results involving NaNs are themselves NaN:

```
>> diff(y)
ans  =
  Columns 1 through 12
      1    1    1   NaN   NaN   1  1  1  1  1  1  1
  Columns 13 through 19
      1  NaN  NaN    1    1    1  1
```

If we calculate the cumulative sum of y, everything from the first NaN onwards is NaN:

```
>> cumsum(y)
ans  =
  Columns 1 through 12
      1    3    6    10  NaN  NaN  NaN  ...  NaN
  Columns 13 through 20
    NaN  NaN  NaN   NaN  NaN  NaN  NaN  NaN
```

MATLAB's surface plotting functions handle NaNs in a similar way:

```
z = peaks;
z(5:35,18:22) = NaN;
subplot(221)
plot(z')
subplot(222)
colormap(gray(64))
imagesc(z)
axis xy
subplot(223)
surfl(z)
shading flat
subplot(224)
contourf(z)
```

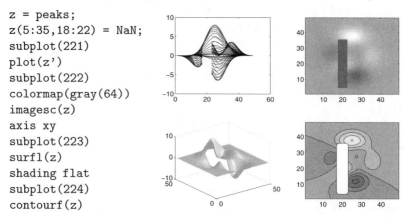

16 Polar Plots

When displaying information which varies as a function of angle, it is often beneficial to use a polar diagram in which conventional (x, y) values are interpreted as angle and radius. Compare the following two displays. First the conventional (x, y) plot:

```
clf
t = linspace(-pi,pi,201);
g = sinc(2.8*sin(t));
plot(t*180/pi,g)
zeroaxes
```

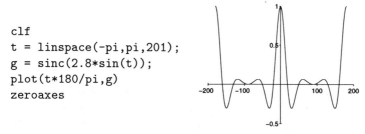

(The command zeroaxes is part of the companion software to this book.) Then the polar diagram indicating the directional variation in the quantity g:

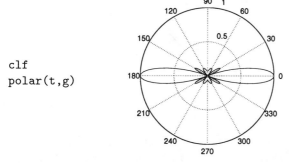

```
clf
polar(t,g)
```

Plots such as these are sometimes displayed in decibel units:

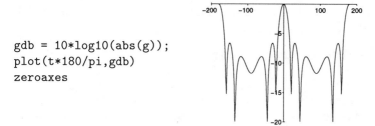

```
gdb = 10*log10(abs(g));
plot(t*180/pi,gdb)
zeroaxes
```

But the polar diagram in this case gives rubbish because it is interpreting the negative decibel values as negative radii:

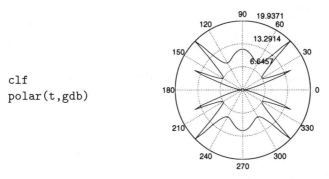

```
clf
polar(t,gdb)
```

In this case you must use a modified version of `polar` that interprets a zero radius as a 0 dB value which should go at the outer limit of the plot. Negative decibel values should appear at smaller radii. I have implemented these ideas in the m-file `negpolar` (see companion software):

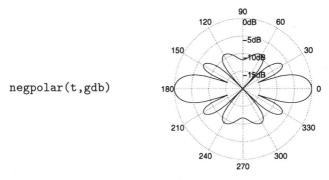

```
negpolar(t,gdb)
```

The `negpolar` function also omits the solid outer circle which, like the box drawn around MATLAB's default `plot` output, unnecessarily frames the plot and can obscure the data that you are trying to display. A faint dotted grid is enough to put the plotted points in context. I will say more about this in the section on Handle Graphics later (see page 65).

17   Fourier Transform

A theorem of mathematics says, roughly, that any function can be represented as a sum of sinusoids of different amplitudes and frequencies. The Fourier transform is the mathematical technique of finding the amplitudes and frequencies of those sinusoids. The Discrete Fourier Transform (DFT) is an algorithm that calculates the Fourier transform for numerical data. The Fast Fourier Transform is an efficient implementation of the DFT. The following functions are available in MATLAB to do Fourier transforms and related operations:

fft	One-dimensional fast Fourier transform
fft2	Two-dimensional fast Fourier transform
fftn	N-dimensional fast Fourier transform
fftshift	Move zeroth lag to centre of transform
ifft	Inverse one-dimensional fast Fourier transform
ifft2	Inverse two-dimensional fast Fourier transform
ifftn	inverse N-dimensional fast Fourier transform
abs	Absolute value (complex magnitude)
angle	Phase angle
cplxpair	Sort complex numbers into complex conjugate pairs
nextpow2	Next power of two
unwrap	Correct phase angles

The FFT of the column vector

```
y = [2 0 1 0 2 1 1 0]';
```

is

```
>> Y = fft(y)
Y =
   7.0000
  -0.7071+ 0.7071i
   2.0000- 1.0000i
   0.7071+ 0.7071i
   5.0000
   0.7071- 0.7071i
   2.0000+ 1.0000i
  -0.7071- 0.7071i
```

The first value of Y is the sum of the elements of y, and is the amplitude of the "zero-frequency", or constant, component of the Fourier series. Terms 2 to 4 are the (complex) amplitudes of the positive frequency Fourier components. Term 5 is the amplitude of the component at the Nyquist frequency, which is half the sampling frequency. The last three terms are the negative frequency components, which, for real signals, are complex conjugates of the positive frequency components.

The fftshift function rearranges a Fourier transform so that the negative and positive frequencies lie either side of the zero frequency.

Companion M-Files Feature 4 *The function fftfreq gives you a two-sided frequency vector for use with fft and fftshift. For example, the frequency vector corresponding to an 8-point FFT assuming a Nyquist frequency of 0.5 is*

```
>> fftfreq(.5,8)'
ans =
```

```
                -0.5000
                -0.3750
                -0.2500
                -0.1250
                      0
                 0.1250
                 0.2500
                 0.3750
```

We combine `fftshift` and `fftfreq` to plot the two-sided FFT:

```
plot(fftfreq(.5,8),fftshift(abs(Y)))
axis([-.5 .5 0 7])
zeroaxes
```

Let us do a slightly more realistic example. We simulate some data recorded at a sampling frequency of 1 kHz, corresponding to a time step dt = 1/1000 of a second. The Nyquist frequency is, therefore, 500 Hz. Suppose there is a 100 Hz sinusoid contaminated by noise. We simulate the data, calculate the FFT, and plot the results as follows:

```
dt = 1/1000;
t = dt:dt:200*dt;
sine = sin(2*pi*100*t);
y = sine + randn(size(t));
Y = fft(y);
f = fftfreq(500,length(Y));
```

```
clf
subplot(211)
stairs(t,y)
hold on
stairs(t,sine-4)
box
xlabel('Time (seconds)')

subplot(212)
stairs(f,fftshift(abs(Y)))
box
xlabel('Frequency (Hz)')
```

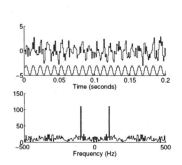

The top trace in the top plot is the noisy data, and the bottom trace is the original pure sinusoid. The lower plot clearly shows the frequency at 100 Hz.

Two GUI-based FFT demos can be accessed by typing `demo` at the prompt. Select the "Signal Processing" option, then choose the "Discrete Fourier Transform" or the "Continuous Fourier Transform".

> **Exercise 5** *Extend the ideas in the previous example to two dimensions, as would be the case, for example, if you made measurements in space and time, rather than time alone. Generate a two-dimensional sinusoid and explore its FFT. (Answer on page 185.)*

18 Power Spectrum

The power spectrum (or power spectral density, or PSD) is a measure of the power contained within frequency intervals. The problem is that we only have a finite set of samples of the true signal so we can never have perfect knowledge about its power spectrum. A common way to estimate a PSD is to use the square of the FFT of the samples. The square of the FFT is called the *periodogram*. The workhorse of MAT-LAB's periodogram-based spectral estimation is the `spectrum` function (in the Signal Processing Toolbox). We illustrate using data similar to the previous example of a noisy sinusoid, but we take more samples. A PSD estimate can be found by typing:

```
dt = 1/1000;
t = dt:dt:8192*dt;
sine = sin(2*pi*100*t);
y = sine + randn(size(t));
clf
spectrum(y)
```

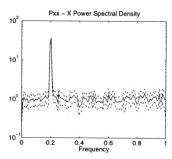

The frequency scale is normalised to the Nyquist frequency. The middle line is the PSD estimate and the two dashed lines are the 95% confidence intervals. Typing `help spectrum` reveals that there are many parameters that you can adjust when calculating the power spectrum. MATLAB's `spectrum` function uses the Welch method of PSD estimation,[6] which divides a long signal into a number of smaller blocks, calculates

---

[6]See Alan V. Oppenheim and Ronald W. Schafer, *Digital Signal Processing*, Prentice-Hall, 1975, p. 553. An excellent general treatment of PSD estimation is also given in William Press, Brian Flannery, Saul Teukolsky and William Vetterling, *Numerical Recipes*, Cambridge University Press, 1989.

the periodograms of the blocks, and averages the periodograms at each frequency. This is a technique commonly used to reduce the variance of the PSD. For example, we can compare the variance of the above estimate to that of a single periodogram by telling spectrum to use a block length equal to the length of the signal:

`spectrum(y,8192)`

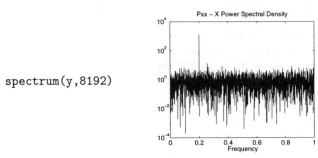

You can also specify windows to reduce spectral leakage, sampling frequencies to get correct frequency scales and overlapping blocks. If you are interested in PSD estimation, the Signal Processing toolbox contains other methods of PSD estimation including Welch's method, MUSIC, maximum entropy and multitaper. MATLAB also provides a graphical user interface for spectral estimation as part of its interactive signal processing environment `sptool`. The System Identification toolbox also contains algorithms for PSD estimation (type `iddemo` and choose option 5 for a demonstration).

19   Sounds in MATLAB

MATLAB can send data to your computer's speaker, allowing you to visually manipulate your data, and listen to it at the same time. A digitised recording of an interesting sound is contained in the mat-file `chirp.mat`. Load this data, do a plot, and listen to the sound by typing:

```
load chirp
plot(y)
sound(y)
```

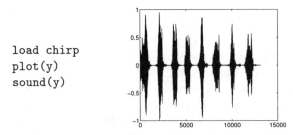

The volume of the sound can be controlled from within MATLAB using the `soundsc` function and supplying an upper and lower limit. Or if you wish, you can use your computer's system software to control the volume. On UNIX the volume of the sound can be controlled with the audiotool. On a PC the volume can be controlled from the "properties" panel of the sound recorder.

You can invoke a sound demo GUI by typing xpsound. This GUI includes these bird chirps plus a few other sounds, three different display types, a volume slider, and a play button.

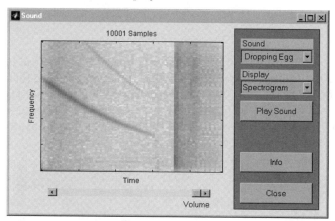

20 Time-Frequency Analysis

Signals, such as the sound data of the previous section, often consist of time series data with a time-varying frequency content. The specgram function allows you to analyse this kind of time-frequency data. As an example we generate a frequency modulated carrier and analyse its frequency variation with time. The modulate and vco function can be used to produce signals with many different modulation types.[7] We begin with a linear frequency sweep from 0 to 500 Hz sampled at 1 kHz. First, you must prepare a frequency control vector, which is normalised between −1 and 1, where −1 corresponds to the minimum frequency and 1 corresponds to the maximum frequency. Here we use a linear frequency control and 8192 points:

```
x = linspace(-1,1,8192);
```

Now use the vco function (in the Signal Processing Toolbox) to convert this to a frequency modulated signal:

```
Fs = 1000;
y = vco(x,[0 500],Fs);
```

The input vector [0 500] says that our frequency sweep will go from 0 Hz to 500 Hz and the sampling frequency is Fs = 1000 Hz. The first thousand points of this signal reveal the steady increase in frequency:

---

[7]In fact what we are doing here could also be done with the m-file chirp.m (not to be confused with the data file chirp.mat).

```
plot(y(1:1000))
axis([0 1000 -5 5])
zeroaxes
```

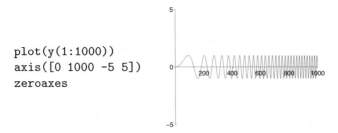

The frequency content of this signal as a function of time can be calculated using the `specgram` function. This function uses the Short Time Fourier Transform (STFT) technique. The STFT chops up the signal into a series of short segments and calculates the FFT of each segment. Each FFT becomes the estimate of the frequency content at that time. For our example we can get a quick display by typing:

```
clf
specgram(y)
colormap(flipud(gray/2+.5))
colorbar
```

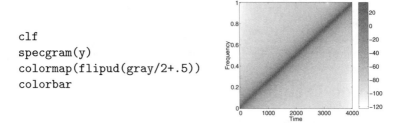

The linear increase in frequency with time is clearly displayed, although here we have not told `specgram` what the sampling frequency is, so it has plotted a normalised frequency scale. If we include the sampling frequency as an input, we get the true frequencies. If you type `help specgram` you will see that the inputs are such that the sampling frequency comes third in the list, after the signal itself and the FFT size. Here we do not want to bother about specifying the FFT size, so we can just specify the empty matrix for that input and `specgram` will use its default value of NFFT = 256:[8]

```
specgram(y,[],Fs)
colormap(flipud(gray/2+.5))
colorbar
```

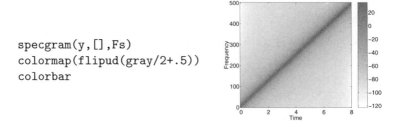

The frequency now goes from zero to 500 Hz.

---

[8]Many of MATLAB's functions behave this way: specifying the empty matrix will tell the function that you want to use its default value for that input.

Exercise 6 *Try a more complicated modulation function; for example, a sinusoidal rather than a linear frequency variation. Try plotting the results as a surface instead of an image. (Answer on page 186.)*

21 Line Animation

MATLAB's comet function can be used to produce an animation on the screen of a trajectory through either two-space or three-space. For example, we use some recorded aircraft GPS data in the file gps.mat.

```
>> clear
>> load gps
>> whos
  Name        Size         Bytes  Class
  t           500x1         4000  double array
  x           500x1         4000  double array
  y           500x1         4000  double array
  z           500x1         4000  double array
Grand total is 2000 elements using 16000 bytes
```

A simple 3-d plot is difficult to interpret:

>> plot3(x,y,z)

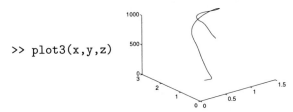

The floating thread has too few visual clues for the eye to interpret, and the altitude variation further clutters the display. A two-dimensional plot tells us that the aircraft was doing turns (but not how high it was):

plot(x,y)
axis equal
box

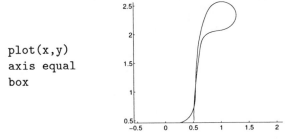

This is an improvement, but we still do not know where the aircraft started, where it finished, and how it went in between. We can see an animation of the trajectory by typing:

```
comet(x,y)
```

(You can get a three-dimensional version by using `comet3`.) You can see it on your screen. But we have just illustrated a disadvantage of such a display: you have to be there. I cannot communicate to you what it looks like on paper. For that you need to resort to, say, an array of two-dimensional plots strung out along the third time dimension. This gets us into the subject of plot arrays, which is discussed in Section 32.3 on page 126.

22 SPTool

SPTool (in the Signal Processing Toolbox) is a graphical user interface to many of MATLAB's signal processing functions. The idea is to import signals from the MATLAB workspace into the SPTool environment where they can be manipulated in a great variety of ways. As an example, load some data into your workspace by typing:

```
load mtlb
```

We will use SPTool to look at this time-series data and calculate various power spectra. Invoke SPTool by typing:

```
sptool
```

Choose the `File→Import` menu item to open the import panel, which allows you to control the variables that `sptool` can "see":

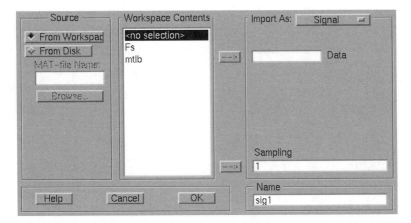

Click on the variable `mtlb` and the arrow button (-->) to get `mtlb` to appear in the `Data` box (or just type `mtlb` there). Do the same to make `Fs` appear in the `Sampling` box. Then press `OK`. A signal called `sig1` appears in the `Signals` box in the main `SPTool` panel. Clicking on the `View` button at the bottom of the `Signals` box opens the signal browser panel:

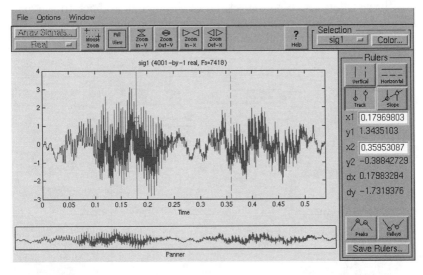

Here you have a plot of the time series with two "rulers". The rulers can be used to pick values out of the data, as well as to calculate intervals and slopes. The data in the `Rulers` box at the right of the display shows this information. At the bottom is a "panner". If you click on the `Zoom In-X` button a couple of times, the top plot shows an expanded portion of the data, and the panner at the bottom shows the location of the top box within the entire signal.

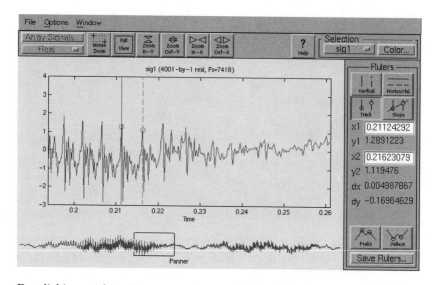

By clicking within the panner box and dragging, you can change the location of the zoomed window. You can listen to this time series by selecting Options→Play.

To calculate the power spectrum of this signal, go back to the main SPTool panel and click the Create button at the bottom of the Spectra box. Doing this will open the Spectrum Viewer:

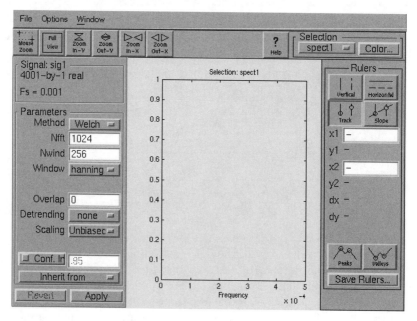

Choose a method with the parameters you like to get a plot of a spectral estimate:

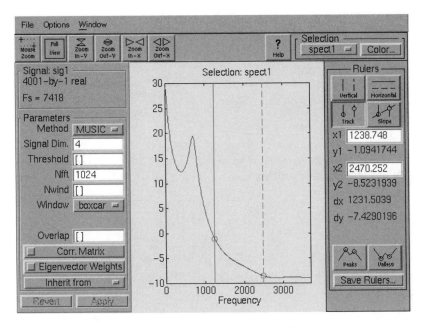

You can design and apply filters to data in a similar way.

23 Handle Graphics

So far in this book, we have only used MATLAB's high-level plotting function (plot, surf, etc.). High-level plotting functions produce simple graphs and automate the many mundane decisions you might make in producing a plot, such as the position of the plot, the colour of the axes, the font size, the line thickness, and so on. MATLAB's system of *Handle Graphics* allows you to control a great many of these "mundane" aspects of plotting, to produce plots that are optimised for communicating the data at hand. The idea behind Handle Graphics is that every object in the figure window (axes, lines, text, surfaces, etc.) has a set of properties. These properties can be examined using the get command and set to new values using the set command. Every object in the figure window also has a unique identifier (a number) called a *handle*. The object's handle tells get and set what object you are interested in. As an introductory example, consider the plot shown on page 58 of the frequency modulated sinusoid:

```
x = linspace(-1,1,8192);
Fs = 1000;
y = vco(x,[0 500],Fs);
plot(y(1:1000))
axis([0 1000 -5 5])
zeroaxes
```

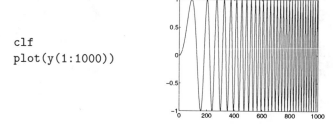

We used the `axis` command to set the y-axis limits to [-5 5] instead
of the default limits, in this case, of [-1 1]

```
clf
plot(y(1:1000))
```

which makes the variation in frequency slightly less apparent, and is
just too grandiose. The eye can pick up very subtle variations in line
straightness, but here the variation is so huge that the lines become
parallel and begin to produce the optical illusion of vibration. Also,
lines that are very nearly vertical or horizontal begin to be affected by
the finite resolution of dot printers. Using Handle Graphics we can
achieve a more elegant result by reducing the height of the y-axis. We
do this by setting the `position` property of the current axes:

```
set(gca,'Position',[.1 .5 .8 .1],'box','off')
```

The gca input is itself a function, which returns the handle to the current
set of axes. We are saying that we want to set the position of the current
axes to be equal to the vector [.1 .1 .8 .1]. The position vector has
the form [*left, bottom, width, height*], in units normalised to the
figure window; $(0,0)$ is the bottom left and $(1,1)$ is the top right. But
perhaps we should shrink it even further, and dispense with the ever-
present axes:

```
set(gca,'Position',[.1 .5 .8 .01],'visible','off')
```

23.1   Custom Plotting Functions

Handle Graphics can be used to write your own graphics m-files that are
fine-tuned to your requirements. For example, the box around the graph
produced by the default `plot` command can obscure the data:

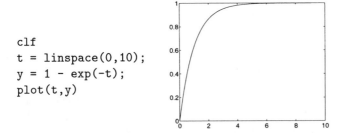

```
clf
t = linspace(0,10);
y = 1 - exp(-t);
plot(t,y)
```

To avoid this problem (which I have found occurs frequently), I use my
own personal version of the `plot` command, called `plt`, which omits the
box:

```
plt(t,y)
```

The m-file for `plt` (see companion software) simply passes all the input
parameters directly to the `plot` command and then sets the `'box'` property of the current plot to `'off'`.

23.2   Set and Get

Typing

> `get(H)`

where H is an object handle, displays all of the property names associated
with the object. Typing

> `set(H)`

displays all of the possible values that can be taken by every property
associated with the object. Typing

> `set(H,'Property')`

displays all of the possible values for the *Property* associated with the
object.

23.3   Graphical Object Hierarchy

MATLAB graphical objects are arranged according to the hierarchy shown here.

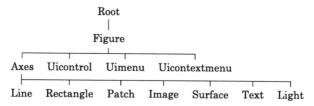

The object immediately above another is called a *parent*, and the objects below are called *children*. In general, children inherit their handle graphics properties from their parent. For example the position of a line on a plot depends on the position of the axes that it goes in, which, in turn, depends on the position of the figure window.

The Root object is the computer screen. There can only be one Root object. You can see the properties of the Root object and the allowable options by typing `set(0)` (the handle of the Root is always equal to zero).

The Uicontrol, Uimenu, and Uicontextmenu objects are graphical user interface elements that are discussed in Part II of this book (see page 133).

A parent can have any number of children. For example the Root can have many Figures, a Figure can have many Axes, and a set of Axes can have many Lines, Surfaces, and so on. If a parent has many children, one of them is designated the *current* one. For example the current set of axes is the one that will be updated the next time you do a `line` command. You can make an object current by clicking on it with the mouse. For example, I clicked on the fourth line from the bottom before setting its `linewidth` property to 5 (the default linewidth is 0.5):

```
plot([1:10]'*[1:10])
set(gco,'linewidth',5)
```

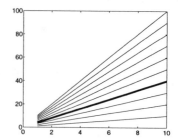

The following functions return the handles of current objects:

gcf Get Current Figure
gca Get Current Axes
gco Get Current Object

The handle of a Figure is the number (1, 2, etc.) that normally appears in the Figure's title bar (supplied by the windowing system).

All of the graphical objects, except the Root object, have low-level creation functions in which you can specify their properties. For example, here is how to create a set of axes with the x-axis tick marks labelled by months of the year:

```
lbls = ['Jan|Feb|Mar|April|May|June|'...
        'July|Aug|Sept|Oct|Nov|Dec'];
clf
axes('position',[.1 .5 .8 .1],'xlim',[1 12],...
     'xtick',1:12,'xticklabel',lbls)
```

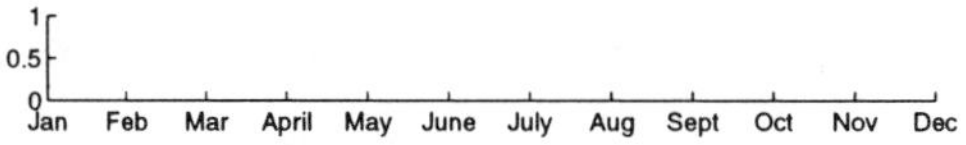

The general format of object creation functions is

$$\text{handle} = function('propertyname','propertyvalue')$$

The output of the function is the handle of the object. This handle can then be used in subsequent calls to `get` and `set` to modify the properties of the object. The *propertyname*s are displayed by MATLAB with capitalisation to make them easier to read; for example, the `VerticalAlignment` text property or the `YAxisLocation` axes property. When you are typing property names, you do not need to use the full name or any capitalisation; you need only use enough letters of the property name to uniquely specify it, and MATLAB does not care what capitalisation you use. Nevertheless, when writing m-files, it is a good idea to use the full property name because abbreviated names may no longer be unique if extra properties are added in future releases of MATLAB.

Example: Line Width

The default way to plot a matrix is to draw one line for each column of the matrix, with the lines differentiated by colour. Suppose instead that we want to differentiate the lines by their thicknesses. One way to do it is as follows. First generate the data and plot it:

```
y = [1:10]'*[1:10];
clf
plot(y)
```

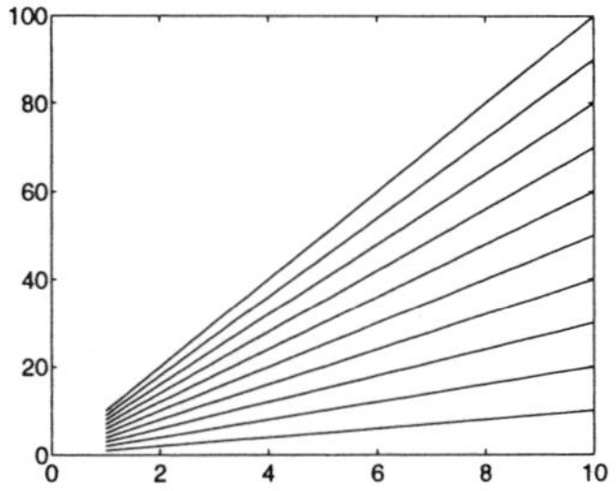

Now we need to get the handles of all the lines. We could have said h = plot(y) to get them, but for now we use the get function:

```
h = get(gca,'children')
```

The gca function returns the handle of the current axes, and get(gca,'children') returns the handles of all the current axes' children (the lines on the plot). Now we want to change the thicknesses of the lines. We set up a vector of line widths with as many elements as there are lines:

```
widths = linspace(.1,10,length(h));
```

The widths of the lines will vary from a minimum of 0.1 to a maximum of 10. We use a for-loop to change the width of each of the lines:

```
for i = 1:10
  set(h(i),'linewidth',widths(i));
end
```

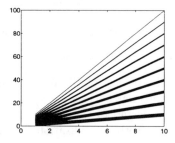

24   Demos

The MATLAB demos are well worth browsing. You can learn about a subject (often reading references are given), as well as learning about MATLAB's capabilities. Of interest to sonar and radar signal processors is MATLAB's Higher Order Spectral Analysis toolbox containing, for example, functions for direction of arrival estimation (beamforming plus other methods), time-frequency distributions, and harmonic estimation. Type help hosa for a list of functions in the Higher Order Spectral Analysis toolbox. Browsing the demos or doing a keyword search may save you from writing your own MATLAB code and re-inventing the wheel. Type demo to get the panel:

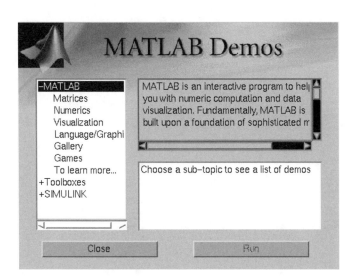

Beyond the Basics

Prelude

This part of the book assumes that you already have some competency with MATLAB. You may have been using it for a while and you find you want to do more with it. Perhaps you have seen what other people do and are wondering how it is done. Well, read on.

This part of the book follows an introductory course in MATLAB (Part I) that covered the basics: matrices, typing shortcuts, basic graphics, basic algebra and data analysis, basics of m-files and data files, and a few simple applications, such as curve fitting, FFTs, and sound. Basic handle graphics were introduced using set and get.

We begin by looking at sparse matrices and strings, go on to deal with some of the data types that are new to MATLAB version 5: cell arrays, multidimensional arrays and structures, then deal with a variety of topics that you will probably have to deal with at some stage if you are a frequent user of MATLAB. The book can be worked through from start to finish, but if you are not interested in a particular topic, you can skip over it without affecting your understanding of later topics. Exercises are given throughout the book, and answers to most of them are given at the end. We start by introducing some new variable types that go beyond the functionality of a rectangular matrix.

25 Sparse Arrays

In some applications, matrices have only a few non-zero elements. Such matrices might arise, for example, when analysing communication networks or when performing finite element modelling. MATLAB provides *sparse arrays* for dealing with such cases. Sparse arrays take up much less storage space and calculation time than full arrays.

25.1 Example: Airfoil

Suppose we are doing some finite element modelling of the airflow over
an aeroplane wing. In finite element modelling you set up a calculation
grid whose points are more densely spaced where the solution has high
gradients. A suitable set of points is contained in the file `airfoil`:

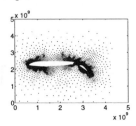

```
load airfoil
clf
plot(x,y,'.')
```

There are 4253 points distributed around the main wing and the two
flaps. In carrying out the calculation, we need to define the network of
interrelationships among the points; that is, which group of points will
be influenced by each point on the grid. We restrict the influence of
a given point to the points nearby. This information is stored in the
vectors i and j, included in the loaded data. Suppose all the points are
numbered $1, 2, \ldots, 4253$. The i and j vectors describe the links between
point i and point j. For example, if we look at the first five elements:

```
>> [i(1:5) j(1:5)]'
ans  =
      1     2     3     5     4
      2     3    10    10    11
```

The interpretation is that point 1 is connected to point 2, point 2 is
connected to point 3, points 3 and 5 are connected to point 10, and so
on. We create a sparse adjacency matrix, A, by using i and j as inputs
to the `sparse` function:

```
A = sparse(i,j,1);
spy(A)
```

The `spy` function plots a sparse matrix with a dot at the positions of
all the non-zero entries, which number 12,289 here (the length of the i
and j vectors). The concentration of non-zero elements near the diagonal
reflects the local nature of the interaction (given a reasonable numbering
scheme). To plot the geometry of the interactions we can use the `gplot`
function:

```
clf
gplot(A,[x y])
axis off
```

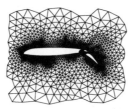

(Try zooming in on this plot by typing `zoom` and dragging the mouse.)
The adjacency matrix here (`A`) is a 4251×4253 sparse matrix with 12,289
non-zero elements, occupying 164 kB of storage. A full matrix of this
size would require 145 MB.

(From now on in this book, the `clf` command will be omitted from
the examples; you will need to supply your own `clf`s where appropriate.)

25.2   Example: Communication Network

Suppose we have a communications network of nodes connected by wires
that we want to represent using sparse matrices. Let us suppose the
nodes are 10 equispaced points around the circumference of a circle.

```
dt = 2*pi/10;
t = dt:dt:10*dt;
x = cos(t)';
y = sin(t)';
plt(x,y)
axis equal off
for i = 1:10
   text(x(i),y(i),int2str(i))
end
```

We want the communications channels to go between each node and its
two second-nearest neighbours, as well as to its diametrically opposite
node. For example, node 1 should connect to nodes 3, 6, and 9; node 2
should connect to nodes 4, 7, and 10; and so on. The function `spdiags`
is used on the following to put the elements of e along the second, fifth,
and eighth diagonals of the (sparse) matrix `A`. If you look at the help for
`spdiags`, you should be able to follow how these statements define the
connection matrix we want. First we define the connection matrix:

```
e = ones(10,1);
A = spdiags(e,2,10,10) + ...
    spdiags(e,5,10,10) + ...
    spdiags(e,8,10,10);
A = A + A';
```

Now do the plot:

```
subplot(221)
spy(A)
subplot(222)
gplot(A,[x y])
axis equal off
for i = 1:10
  text(x(i),y(i),int2str(i))
end
```

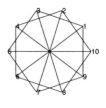

The plots show the sparse node-connection matrix on the left and the physical connection on the right.

> **Exercise 7** *Repeat this communications example for the case of 100 nodes around the circle. Then try changing the connection matrix. (Answer on page 187.)*

26   Text Strings

A string is an array of characters. For example, these are strings: 'hello', 'John Smith', and '12'. The last one is a string, not the number 12, because it is surrounded by quotes. MATLAB represents characters as their ASCII values. You can convert between ASCII values and the characters they represent using the **double** and **char** commands:

```
>> alph = 'ABCDE'
alph  =
ABCDE

>> num = double(alph)
num  =
    65    66    67    68    69

>> char(num)
ans  =
ABCDE

>> char(num+5)
ans  =
FGHIJ
```

The **double** command converts its argument to double precision values, the default MATLAB format.

To get a quote character in a string use two quotes in succession:

```
>> str = 'you''re the one'
```

```
str  =
you're the one
>> str = '''you''re the one'''
str  =
'you're the one'
```

> **Exercise 8** *Create a table of integers from 0 to 255 and their equivalent* ASCII *values. Printing which* ASCII *"character" rings the system bell? (Answer on page 187.)*

26.1 String Matrices

To create a matrix of strings, use the semicolon to separate the lines:

```
>> m = [alph ; char(num+5) ; 'KLMNO']
m  =
ABCDE
FGHIJ
KLMNO
```

You cannot create a matrix of strings having different lengths:

```
>> z = [alph ; 'b']
??? All rows in the bracketed expression must have the
same number of columns.
```

(You should use cell arrays—discussed later—if you really want to create a "matrix" like this.) To simulate the effect, though, you can pad with zeros:

```
>> z = ['abcd' ; 'b   ']
z  =
abcd
b
```

The second line has three blank spaces to the right of the "b". A convenient way to do this is to use the char function, which does the padding for you:

```
>> z = char('These','lines are','of varying lengths.')
z  =
These
lines are
of varying lengths.
```

26.2   Comparing Strings

The = = test is not a good idea with strings because it compares the ASCII values of the strings, which must have the same length; if the strings are not the same length, you get an error. The `strcmp` command avoids this difficulty:

```
>> c1 = 'blond';
>> c2 = 'brown';
>> c3 = 'blonde';
>> c1 = = c2
ans =
     1    0    1    0    0
>> c2 = = c3
??? Array dimensions must match for binary array op.

>> strcmp(c2,c3)
ans =
     0
```

26.3   String Manipulations

Typing `help strfun` displays the full set of commands for working with strings. A common example is to identify words within a string by searching for whitespace (blank characters, tabs, etc.):

```
>> str = 'I go now';
>> isspace(str)
ans =
     0    1    0    0    1    0    0    0
```

You can also search for letters:

```
>> isletter(str)
ans =
     1    0    1    1    0    1    1    1
```

To find where a shorter string occurs within a longer one, use the `findstr` command:

```
>> pos = findstr(str,'go')
pos =
     3
>> pos = findstr(str,'o')
pos =
     4    7
```

To replace one string with another, use the `strrep` command:

```
>> strrep(str,'go','am')
ans  =
I am now
```

The replacement text need not be the same length as the text it is replacing:

```
>> strrep(str,'go','eat snails')
ans  =
I eat snails now
```

And the text to be replaced can occur more than once:

```
>> strrep(str,'o','e')
ans  =
I ge new
```

To delete characters from a string, replace them with an empty string '' or []:

```
>> strrep(str,'o','')
ans  =
I g nw
```

26.4 Converting Numbers to Strings

The functions num2str and int2str are useful for general purpose conversion of numbers to strings. The latter is for integers:

```
>> for i = 1:3
disp(['Doing loop number ' int2str(i) ' of 3'])
end
Doing loop number 1 of 3
Doing loop number 2 of 3
Doing loop number 3 of 3
```

And num2str is for everything else:

```
>> for i = 1:3
disp(['Case ' int2str(i) ', sin = ' num2str(sqrt(i))])
end
Case 1, sin = 1
Case 2, sin = 1.4142
Case 3, sin = 1.7321
```

The inputs can be vectors or matrices:

```
>> v = sin((1:3)*pi/6)
v   =
     0.5000    0.8660    1.0000
>> num2str(v)
ans   =
0.5    0.86603    1
>> q = reshape(1:9,3,3)
q   =
     1     4     7
     2     5     8
     3     6     9

>> int2str(q)
ans   =
1  4  7
2  5  8
3  6  9
>> size(ans)
ans   =
     3     7
```

You can tell num2str how many digits to display by giving it a second
parameter:

```
>> num2str(pi,2)
ans   =
3.1
>> num2str(pi,15)
ans   =
3.14159265358979
```

The second parameter of num2str can also specify the format by means
of C language conversions. These involve the percent character, width
and precision fields, and conversion characters: d, f, e, etc. (see table
below). The basic idea is to use a string of characters beginning with %
to control the formatting. For example, to output five decimal places in
a field of 12 characters with exponential notation, use:

```
>> num2str(pi,'%12.5e')
ans   =
3.14159e+00

>> num2str(-pi,'%12.5e')
ans   =
-3.14159e+00

>> num2str(pi*1e100,'%12.5e')
```

```
ans  =
3.14159e+100
```

Some additional text can be mixed with the numerical conversion, for example:

```
>> num2str(pi,'Pi has a value of %12.5e, or thereabouts.')
ans  =
Pi has a value of  3.14159e+00, or thereabouts.
```

The online help[9] entry for the sprintf command gives a full description of how to use the various formatting options. (The sprintf command is the MATLAB version of the C language command of the same name.) The following table is taken from the online help.

%c	Single character
%d	Decimal notation (signed)
%e	Exponential notation (using a lowercase e as in 3.1415e+00)
%E	Exponential notation (using an uppercase E as in 3.1415E+00)
%f	Fixed-point notation
%g	The more compact of %e or %f. Insignificant zeros do not print.
%G	Same as %g, but using an uppercase E
%o	Octal notation (unsigned)
%s	String of characters
%u	Decimal notation (unsigned)
%x	Hexadecimal notation (using lowercase letters a-f)
%X	Hexadecimal notation (using uppercase letters A-F)

To further control the formatting, other characters can be inserted into the conversion specifier between the % and the conversion character:

Character	What it does
A minus sign (-)	Left-justifies the converted argument in its field.
A plus sign (+)	Always prints a sign character (+ or −).
Zero (0)	Pads with zeros rather than spaces.
Digits (field width)	Specifies the minimum number of digits to be printed.
Digits (precision)	Specifies the number of digits to be printed to the right of the decimal point.

---

[9]Type helpdesk at the command line to get hypertext help.

Examples:

`sprintf('%0.5g',(1+sqrt(5))/2)`	`1.618`
`sprintf('%0.5g',1/eps)`	`4.5036e+15`
`sprintf('%10.3f',-pi)`	`-3.142`
`sprintf('%10.3f',-pi*1000000)`	`-3141592.654`
`sprintf('%10.3f',-pi/1000000)`	`-0.000`
`sprintf('%d',round(pi))`	`3`
`sprintf('%s','hello')`	`hello`
`sprintf('The array is %dx%d.',2,3)`	`The array is 2x3.`
`sprintf('\n')`	Line termination character on all platforms

These functions are "vectorised", meaning that if you input a non-scalar, then all the elements will be converted:

```
>> str = num2str(rand(3,3),6)
str =
0.502813    0.304617    0.682223
0.709471    0.189654    0.302764
0.428892    0.193431    0.541674
>> size(str)
ans =
     3    32
```

Exercise 9 *Explore the operation of the following m-file that breaks a sentence up into a list of words.*

```
function all_words = words(input_string)
remainder = input_string;
all_words = '';
while any(remainder)
  [chopped,remainder] = strtok(remainder);
  all_words = strvcat(all_words,chopped);
end
```

Why do you think `strvcat` *is used instead of* `char`*? (Answer on page 188.)*

26.5   Using Strings as Commands

The `eval` Function

The `eval` function takes a string input and executes it as a MATLAB command. For example:

```
>> str = 'v = 1:5'
str =
v = 1:5
```

```
>> eval(str)
v  =
      1     2     3     4     5
```

The eval(str) statement acts just as if we had typed v = 1:5 at the command line. To suppress the output we need to add a semicolon character to the end of the string:

```
>> str = 'v = 1:5;'
str  =
v = 1:5;
>> eval(str)
```

The eval command now produces no output, while still defining the variable v. To take another example, let us suppose we want to define a set of vectors $v_i = 1, 2, \ldots i$ for $i = 1, 2 \ldots 10$. At the command line we could type:

```
v1 = 1;
v2 = 1:2;
v3 = 1:3;
```

and so on. The eval command provides a neat solution:

```
>> clear
>> for i = 1:10
      str = ['v' int2str(i) ' = 1:i;'];
      eval(str)
   end
```

This has generated the variables v1, ..., v10, with the appropriate elements:

```
>> whos
  Name       Size         Bytes  Class
  i          1x1              8  double array
  str        1x10            20  char array
  v1         1x1              8  double array
  v10        1x10            80  double array
  v2         1x2             16  double array
  v3         1x3             24  double array
  v4         1x4             32  double array
  v5         1x5             40  double array
  v6         1x6             48  double array
  v7         1x7             56  double array
  v8         1x8             64  double array
  v9         1x9             72  double array
Grand total is 66 elements using 468 bytes
```

```
>> v6
v6  =
     1     2     3     4     5     6
```

The feval Function

The **feval** command is like **eval**, except that it is used for evaluating named functions. An example would be:

```
str = 'sin';
t = linspace(0,2*pi);
q = feval(str,t);
plt(t,q)
```

If **str** is a string containing the name of a function, then **y = feval(str,x)** evaluates that function for the input argument **x**. Another example defines data for plotting by looping over the trigono-metric functions **sin**, **cos**, and **tan** contained within a single matrix of characters (the command **zeroaxes** is part of the companion software to this book):

```
str = ['sin';'cos';'tan'];
for i = 1:3
    q(i,:) = feval(str(i,:),t);
end
clf
plt(t,q)
axis([0 2*pi -6 6])
zeroaxes
```

Inline Objects

Inline objects allow you to store a function as a string and use it much as you would write it symbolically. This, for example, is how to define the parabola $f(x) = (x+1)(x-1)$:

```
>> f = inline('(x + 1).*(x - 1)')
f  =
     Inline function:
     f(x) = (x + 1).*(x - 1)
```

We can now evaluate $f(3)$ by typing:

```
>> f(3)
ans  =
     8
```

Inline objects, like every other MATLAB construction, is vectorised:

```
>> f(0:4)
ans =
    -1    0    3    8   15
```

They can be used in place of other variables:

```
x = linspace(-4,4);
clf
plt(x,f(x))
hold on
plt(x,f(x-2),'--')
zeroaxes
```
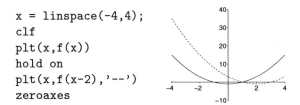

Exercise 10 *Create a function called funplot that takes the name of a function and a range of x-values and produces a plot over that range. For example, the following input should produce this plot:*

funplot('sin',[0 pi])
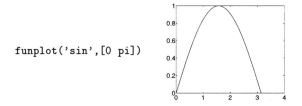

The function should work whether you type funplot('sin',[0 pi]) or funplot('sin(x)',[0 pi]). Hint: What are the ASCII values of (and)? (Answer on page 188.)

27 Cell Arrays

Cell arrays are arrays of different things. The "things" can be scalars, vectors, matrices, strings (of different length), structures (see section on structures), or other cell arrays. For example, when we looked at string matrices we saw that we had to pad the rows with blanks to make them all the same length. Using a cell array, we can create a "ragged-right matrix":

```
t = {'O sacred receptacle of my joys,';
'Sweet cell of virtue and nobility,';
'How many sons of mine hast thou in store,';
'That thou wilt never render to me more!'}
```

The curly brackets { and } denote cells. The cell we created above is a 4×1 cell array:

```
>> whos
  Name       Size           Bytes  Class
    t         4x1             658  cell array
Grand total is 149 elements using 658 bytes
>> t(1)
ans =
    'O sacred receptacle of my joys,'
>> t{1}
ans   =
O sacred receptacle of my joys,
>> t{1}(1)
ans   =
O
>> t{1}(1:8)
ans   =
O sacred
```

Let us add another element to the cell array by putting a 3×3 matrix in the first row of the second column:

```
>> t{1,2} = spiral(3)
t =
    [1x31 char]     [3x3 double]
    [1x34 char]             []
    [1x41 char]             []
    [1x39 char]             []
```

MATLAB has filled the rest of the cells in column 2 with empty cells. We used the curly brackets t{1,2} to refer to that particular cell. If we had used ordinary round brackets, we would have produced an error:

```
>> t(1,2) = spiral(3)
??? Conversion to cell from double is not possible.
```

This is because there is a difference between indexing *cells* and indexing their *contents*. For example, to extract the word "virtue" from the second line of the quotation in the first column, we need to access the *cell* {2,1}, then get characters 15 to 20 from that cell's *contents*:

```
>> t{2,1}(15:20)
ans   =
virtue
```

When assigning a cell you can use the curly brackets on either the left or right hand side of the equals sign, but you must put them *somewhere*, to tell MATLAB that you want this to be a cell. Otherwise, MATLAB thinks you are defining a mathematical matrix and gives you an error to the effect that the things on each side of the equal sign have different sizes. For example, we can type:

```
>> a(1,1) = {[1 2 3]}
a =
    [1x3 double]
```

or

```
>> clear a
>> a{1,1} = [1 2 3]
a =
    [1x3 double]
```

but not

```
>> clear a
>> a(1,1) = [1 2 3]
??? In an assignment  A(matrix,matrix) = B, the number of
columns in B and the number of elements in the A column
index matrix must be the same.
```

Cell arrays can contain other cell arrays. For example:

```
>> t = {'Fred Flintstone' {[1 2 3] , spiral(3)}}
t =
    'Fred Flintstone'    {1x2 cell}
```

MATLAB's default display of a cell array is in summary form, as in the above examples. You can display the details using celldisp:

```
>> celldisp(t)
t{1}  =
Fred Flintstone
t{2}{1}  =
    1    2    3
t{2}{2}  =
    7    8    9
    6    1    2
    5    4    3
```

Or, you can get a graphical summary using cellplot:

cellplot(t)

The left-hand box is the first cell, containing the string 'Fred Flintstone'. The right-hand box is the second cell containing a 1×2 cell array whose cells contain the vector [1 2 3] and the matrix spiral(3), respectively.

To index nested cell arrays, use as many sets of curly brackets { and } as needed to get you to the level of nesting required, then use round brackets (and) to access their contents. For example:

```
>> tt = {t  {'Barney Rubble' {[-1 1] , 'Bedrock'}}}
tt =
    {1x2 cell}
    {1x2 cell}
>> cellplot(tt)
```

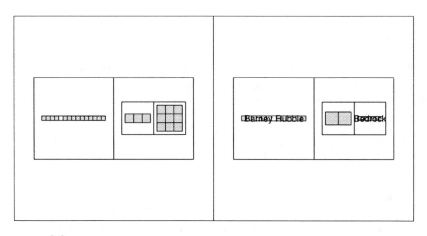

```
>> tt{2}
ans =
    'Barney Rubble'      {1x2 cell}
>> tt{2}{2}
ans =
    [1x2 double]      'Bedrock'
>> tt{2}{2}{1}
ans  =
    -1      1
>>  tt{2}{2}{2}
ans  =
Bedrock
>> tt{2}{2}{1}{2}
??? Cell contents reference from a non-cell array object.

>> tt{2}{2}{1}(2)
ans  =
    1
```

Exercise 11 *Do you know where the word "stuck" has come from in the following example (answer on page 189):*

```
>> t = {'help' spiral(3) ; eye(2) 'I''m stuck'};
>> tt = {t t ;t' fliplr(t)};
>> tt{2,2}{2,1}(5:9)
ans =
stuck
```

28 Structures

Structures are arrays whose names have dot-separated parts. They can be used to store information of different kinds together in a hierarchical structure. Let us do a simple example:

```
>> staff.name = 'John Smith'
staff =
    name: 'John Smith'
>> staff.age = 43
staff =
    name: 'John Smith'
     age: 43
>> staff.favourites = [1 42 37]
staff =
              name: 'John Smith'
               age: 43
        favourites: [1 42 37]
```

We have created a structure called staff which is of size 1×1:

```
>> whos
  Name        Size        Bytes  Class
  staff       1x1           424  struct array
```

The staff structure has three fields: name, age, and favourites:

```
>> staff
staff =
              name: 'John Smith'
               age: 43
        favourites: [1 42 37]
```

To add another staff member's data to this structure, add subscripts to define a second element:

```
staff(2).name = 'Jane Smythe';
staff(2).age = 30;
staff(2).favourites = [pi eps realmax realmin NaN Inf];
```

The sizes of the fields do not have to be the same for each element of the structure. For example, Jane Smythe's favourite vector contains more elements than John Smith's.

28.1   Example: Meteorological Database

In this example we create a structure using the `struct` function. We create a meteorological observation database as follows:

```
meteo = struct('Site',{'Adelaide','Sydney'},...
               'Time',{2.3 4},...
               'Temperature',{24 19},...
               'Pressure',{1023 1015})
```

This structure consists of temperature and pressure measurements at two different times at Sydney and Adelaide. The Adelaide data was taken at 2:30:

```
>> meteo(1)
ans =
            Site: 'Adelaide'
            Time: 2.3000
     Temperature: 24
        Pressure: 1023
```

and the Sydney data was taken at 4:00:

```
>> meteo(2)
ans =
            Site: 'Sydney'
            Time: 4
     Temperature: 19
        Pressure: 1015
```

Let us suppose we have some new Sydney data taken at 8:00 and 11:00. We add this as follows:

```
>> meteo(2).Time(2:3) = [8 11];
>> meteo(2).Temperature(2:3) = [16.5 15.3]
meteo =
1x2 struct array with fields:
    Site
    Time
    Temperature
    Pressure
>> meteo(2).Temperature
ans   =
    19.0000    16.5000    15.3000
```

The pressure meter broke so we do not have new pressure data for these two new times. We could leave the pressure field with one entry, but it might be better to indicate the absence of data more explicitly with NaNs:

```
>> meteo(2).Pressure(2:3) = [NaN NaN];
>> meteo(2).Pressure
ans  =
        1015            NaN            NaN
```

Suppose now that we have discovered that all our pressure readings were
wrong; we need to delete the pressure field altogether from the structure:

```
>> meteo = rmfield(meteo,'Pressure')
meteo =
1x2 struct array with fields:
    Site
    Time
    Temperature
```

Now we want to add humidity measurements at the two sites. Suppose
Adelaide's humidity was 69% and Sydney's was 86%, 80%, and 76% at
the three different times:

```
[meteo.humidity] = deal(69,[86 80 76]);
>> meteo(1)
ans =
            Site: 'Adelaide'
            Time: 2.3000
     Temperature: 24
        humidity: 69
>> meteo(2)
ans =
            Site: 'Sydney'
            Time: [4 8 11]
     Temperature: [19 16.5000 15.3000]
        humidity: [86 80 76]
```

(The `deal` command copies a list of inputs to a list of outputs.)
  To do operations on field elements, just treat them as any other
MATLAB array:

```
>> meteo(2).Temperature
ans  =
   19.0000   16.5000   15.3000
>> mean(meteo(2).Temperature)
ans  =
   16.9333
```

The temperature measurements at both sites in the structure are
accessed by typing:

```
>> meteo.Temperature
ans  =
    24
ans  =
   19.0000   16.5000   15.3000
```

We can capture this output in a cell array as follows:

```
>> q = {meteo.Temperature}
q =
    [24]    [1x3 double]
```

Or, we can string them together in a single array by enclosing the
meteo.Temperature expression in square brackets:

```
>> q = [meteo.Temperature]
q  =
   24.0000   19.0000   16.5000   15.3000
```

In this way you can operate on all elements of a field at once. For
example, to calculate the mean of all the temperature measurements:

```
>> mean([meteo.Temperature])
ans  =
   18.7000
```

28.2   Example: Capturing the List of Variables

Typing whos gives you a list of the variables present in the workspace,
along with their size, the number of bytes they occupy, and their class.
For example, create the following variables:

```
clear
a = 1;
name = 'Jane Smythe';
vect = [1 2 3];
acell = {1 2 ; 'big' 'little'};
meteo = struct('Site',{'Adelaide','Sydney'});
```

The whos command produces the following list:

```
>> whos
  Name        Size          Bytes  Class
  a           1x1               8  double array
  acell       2x2             402  cell array
  meteo       1x2             244  struct array
  name        1x11             22  char array
  vect        1x3              24  double array
Grand total is 58 elements using 1396 bytes
```

We can capture this list by giving `whos` an output variable:

```
>> varlist = whos
varlist =
5x1 struct array with fields:
    name
    size
    bytes
    class
```

The average size of the variables is

```
>> mean([varlist.bytes])
ans  =
  140
```

A cell array of variable names can be generated by:

```
>> names = {varlist.name}
names =
    'a'     'acell'     'meteo'     'name'     'vect'
```

Similar structures are generated by giving output arguments to `what` and `dir`.

29   Multidimensional Arrays

Multidimensional matrices are natural extensions of the normal two-dimensional matrices for cases where the data represent more than two dimensions. Examples are

- Medical tomography, where three-dimensional volumetric data are built up from a series of two-dimensional images;

- Temperature measurements taken at a three-dimensional grid in a room;

- Temperature measurements taken at a three-dimensional grid in a room and at a sequence of times, leading to a four-dimensional data set;

- Red, green and blue components of a two-dimensional image, an $M \times N \times 3$ matrix; and

- Acoustic measurements of sound spectra as a function of frequency, direction of arrival, and time (sonar).

Let us get the hang of things by generating a $3 \times 3 \times 3$ matrix:

```
>> a = [1 2 3;4 5 6;7 8 9]
a   =
       1       2       3
       4       5       6
       7       8       9
>> a(:,:,2) = a*2
a(:,:,1)   =
       1       2       3
       4       5       6
       7       8       9
a(:,:,2)   =
       2       4       6
       8      10      12
      14      16      18
>> a(:,:,3) = eye(3)
a(:,:,1)   =
       1       2       3
       4       5       6
       7       8       9
a(:,:,2)   =
       2       4       6
       8      10      12
      14      16      18
a(:,:,3)   =
       1       0       0
       0       1       0
       0       0       1
```

Multidimensional arrays must be full *N*-rectangles; that is, they must have the same number of elements in parallel dimensions: all rows must have the same number of columns, all "pages" must have the same number of rows and columns, etc.

If you assign a single value to a matrix, MATLAB expands the definition as you would expect:

```
>> a(:,:,3) = 3
a(:,:,1)   =
       1       2       3
       4       5       6
       7       8       9
a(:,:,2)   =
       2       4       6
       8      10      12
      14      16      18
```

```
a(:,:,3)  =
      3      3      3
      3      3      3
      3      3      3
```

Indexing for multidimensional arrays works in the same way as two-dimensional arrays;

```
>> a(2,:,1)
ans  =
      4      5      6
>> a(2,:,2)
ans  =
      8     10     12
>> a(2,:,:)
ans(:,:,1)  =
      4      5      6
ans(:,:,2)  =
      8     10     12
ans(:,:,3)  =
      3      3      3
```

Data can be removed from multidimensional arrays by using the empty matrix:

```
>> a(:,:,2) = []
a(:,:,1)  =
      1      2      3
      4      5      6
      7      8      9
a(:,:,2)  =
      3      3      3
      3      3      3
      3      3      3
```

Elements can be columnarly extracted from multidimensional arrays in the same way as they are from two-dimensional arrays:

```
>> a(:)'
ans  =
  Columns 1 through 12
      1    4    7    2    5    8    3    6    9    3    3    3
  Columns 13 through 18
      3    3    3    3    3    3
```

29.1   Generating Multidimensional Grids

The function `meshgrid` can be used to create matrices representing evenly-spaced grids of points.

```
>> [x,y] = meshgrid(1:5,1:3)
x   =
      1     2     3     4     5
      1     2     3     4     5
      1     2     3     4     5
y   =
      1     1     1     1     1
      2     2     2     2     2
      3     3     3     3     3
>> clf
>> plt(x,y,'o')
>> axis([0.9 5 0.9 3])
```

Such matrices can be used, for example, as variables in functions of x and y:

```
[x,y] = meshgrid(linspace(0,5),linspace(-10,10));
r = sqrt(x.^2 + y.^2);
contour(x,y,r)
axis equal
axis([-10 10 -10 10])
```

(More detail on the `axis` command can be found on page 119.) The `meshgrid` function can be used to produce three-dimensional grids, returning three-dimensional arrays that can be used in an analogous manner. To go to more than three dimensions, you can use the function `ndgrid`. The following example of a three-dimensional volume visualisation is taken from the help entry for `ndgrid`:

```
[x1,x2,x3] = ndgrid(-2:.2:2, -2:.25:2, -2:.16:2);
z = x2 .* exp(-x1.^2 - x2.^2 - x3.^2);
```

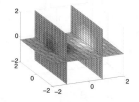

```
slice(x2,x1,x3,z,[-1.2 .8],[], -.2)
view(-24,28)
```

> **Exercise 12** *What is the difference between the outputs of* meshgrid *and* ndgrid *when generating grid matrices of less than four dimensions? Why this difference? (Answer on page 189.)*

29.2 Operations with Multidimensional Arrays

Many matrix operators work with multidimensional arrays. For example, the columnar sum of our $3 \times 3 \times 2$ matrix, a, is

```
>> a
a(:,:,1)   =
       1       2       3
       4       5       6
       7       8       9
a(:,:,2)   =
       3       3       3
       3       3       3
       3       3       3
>> sum(a)
ans(:,:,1)   =
      12      15      18
ans(:,:,2)   =
       9       9       9
```

If you look carefully, you will see that the result of the sum is a $1 \times 3 \times 2$ matrix:

```
>> size(sum(a))
ans   =
       1       3       2
```

This is not the same as a 3×2 matrix. If you want the result to be a 3×2 matrix, you can use the **squeeze** function, which gets rid of singleton dimensions:

```
>> squeeze(sum(a))
ans   =
      12       9
      15       9
      18       9
```

MATLAB does not do an automatic squeeze whenever the result has singleton dimensions because there are times when you need the singleton dimension to add more data.

If you want to sum over other dimensions than the rows, you give a second parameter to the sum function specifying the dimension you want to sum over. For example, to sum over columns:

```
>> sum(a,2)
ans(:,:,1)   =
       6
      15
      24
```

```
ans(:,:,2) =
     9
     9
     9
```

And to sum over "pages":

```
>> sum(a,3)
ans =
      4     5     6
      7     8     9
     10    11    12
```

Note that sum(a) is equal to sum(a,1). The sum over "pages" gives a 3×3 matrix, which is the same as a $3 \times 3 \times 1$ matrix.

The sum function and other functions that operate on vectors, like mean, diff, max, and so on, work as you might expect them to for multi-dimensional arrays. By default they usually operate on the first non-singleton dimension of the array. Many functions that operate on two-dimensional matrices do not have such straightforward multidimensional extensions. For example, if we try to take the transpose of our matrix:

```
>> a'
??? Error using = = > '
Transpose on ND array is not defined.
```

The transpose operation (exchanging rows and columns) makes no sense here because it is insufficiently specified. (If you want to rearrange a multidimensional array's dimensional ordering, use the permute function; in our example, try permute(a,[2 1 3])). Another example is the eigenvalue operator eig, which has no mathematical meaning for multidimensional arrays. In fact, none of the functions that appear if you type help matfun has a reasonable meaning for multidimensional matrices. Nor do the matrix operators *, ^, \ or /.

29.3   RGB Images

Introduction to RGB Images

RGB images in MATLAB are $M \times N \times 3$ matrices consisting of red, green, and blue intensity maps. When such a three-dimensional matrix is used as an input to the image command, MATLAB adds the red, green, and blue intensities to give the right colours on the screen. To illustrate the idea, our first example reproduces three overlapped discs of red, green, and blue light to give yellow, cyan, magenta, and white overlaps. We generate matrices of (x,y) points covering the plane from -2 to 2:

```
[x,y] = meshgrid(linspace(-2,2,200));
```

We define a red disc by setting all the pixels that are within a circle to one; all the other pixels are zero. The circle is defined by the equation:

$$(x - x_0)^2 + (y - y_0)^2 = R^2 ,$$

where (x_0, y_0) is the centre of the circle, and R is the radius. We set the centre of the red disc to $(-0.4, -0.4)$ and the radius to 1.0:

```
R = 1.0;
r = zeros(size(x));
rind = find((x + 0.4).^2 + (y + 0.4).^2 < R^2);
r(rind) = 1;
```

The green and blue discs are defined in the same way, just shifting the centre of the circle in each case:

```
g = zeros(size(x));
gind = find((x - 0.4).^2 + (y + 0.4).^2 < R^2);
g(gind) = 1;
b = zeros(size(x));
bind = find(x.^2 + (y - 0.4).^2 < R^2);
b(bind) = 1;
```

Now we concatenate the matrices r, g, and b into one $200 \times 200 \times 3$ matrix called rgb:

```
rgb = cat(3,r,g,b);
```

We use rgb as an input to imagesc, which interprets the intensities in the range 0.0 to 1.0:

```
imagesc(rgb)
axis equal off
```

On your screen you can see these as overlapped discs of coloured light.

> **Exercise 13** *Redefine the red, green, and blue discs so that instead of a circular disc of light at uniform maximum intensity, the intensity increases within each circle from zero at the centre to one at the edge; outside the circles the intensity should be zero. Create the new overlapped image. (Answer on page 189.)*

An Application of RGB Images

To see how RGB images can be used, we look at how an image can be filtered. An image of the Cat's Eye Nebula, which is stored on disk as a JPEG image, can be read into MATLAB using the `imread` command:

```
>> q = imread('ngc6543a.jpg');
>> size(q)
ans =
    650   600     3
```

The result is a $650 \times 600 \times 3$ matrix, where the "pages" represent respective red, green, and blue intensities. We can display the image by typing:

```
image(q)
axis image off
```

(See page 120 for a description of `axis image`.) On your screen this appears as a colour image. Suppose we want to filter out the red component. We do this by setting the first "page", the red component of the image, equal to zero. First we take a copy of the original so we can later plot the two images side by side:

```
q_original = q;
q(:,:,1) = 0;
subplot(221)
image(q_original)
axis image off
subplot(222)
image(q)
axis image off
```

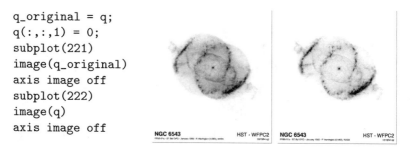

29.4   Example: Sonar

Let us look at some sonar data consisting of sound spectral power levels measured as a function of frequency, direction of arrival, and time. Load this data from the file `sonar.mat`:

```
>> load sonar
```

```
>> whos
  Name         Size          Bytes  Class
  data         128x103x9    949248  double array
  f            1x103           824  double array
  t            1x9              72  double array
  th           1x128          1024  double array
```

The data consists of spectra measured at 103 frequencies, 128 arrival angles, and 9 time steps. Let us plot the fifth time sample:

```
colormap(flipud(gray))
imagesc(f,th,data(:,:,5))
axis xy
colorbar
xlabel('Frequency, Hz')
ylabel('Arrival angle, degrees')
```

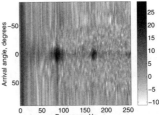

Darker colours correspond to higher intensities. You can see two strong sources at an angle of zero degrees and at frequencies of 85 and 170 Hz. The fact that $170 = 2 \times 85$ might lead us to suspect that the 170 Hz source is just the first harmonic of the 85 Hz source. Let us look at all of the time samples together. This time we'll cut off the lower intensities by setting the minimum colour to correspond to an intensity of 5 (this is the call to caxis in the following code). We also turn off the y-axis tick labels for all but the first plot, and we make the tick marks point outwards:[10]

```
for i = 1:9
  subplot(3,9,i), imagesc(f,th,data(:,:,i)), axis xy
  set(gca,'tickdir','out')
  if i == 1
    ylabel('Arrival angle, degrees')
    xlabel('Frequency, Hz')
  end
  if i>1, set(gca,'yticklabel',[]), end
  caxis([5 Inf]), title(['i_t = ' num2str(t(i))])
end
```

---

[10]See Handle Graphics Sections 23 and 31 (pages 63 and 107).

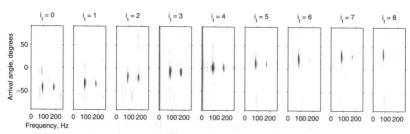

You can see that over the 8 time steps, the arrival direction of the sound
has changed from −45 degrees to 30 degrees, and the two sources always
come from the same direction, strengthening our notion that the two are
in fact harmonics of the same source. Let us look at the time-frequency
and time-angle distributions of this data. The above output of the whos
command shows that the row index of data corresponds to the different
angles, so if we calculate the mean over the rows we will be left with a
time-frequency distribution:

```
>> time_freq = mean(data);
>> size(time_freq)
ans  =
      1    103      9
```

We are left with a $1 \times 103 \times 9$ matrix of averages over the 128 arrival
angles. To plot the results we have to squeeze this to a two-dimensional
103×9 matrix:

```
time_freq = squeeze(mean(data));
imagesc(t,f,time_freq)
axis xy
xlabel('Time, s')
ylabel('Frequency, Hz')
```

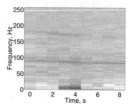

The frequency varies slightly with time. By averaging the rows of the
data matrix we can get a similar plot of the variation of arrival angle
with time:

```
time_angle = squeeze(mean(data,2));
imagesc(t,th,time_angle)
axis xy
xlabel('Time, s')
ylabel('Arrival angle, degrees')
```

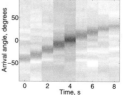

29.5   Multidimensional Cell Arrays

Multidimensional cell arrays are just like ordinary multidimensional
arrays, except that the cells can contain not only numbers, but vectors,

matrices, strings, structures, or other cell arrays. For example, to create a $2 \times 2 \times 2$ cell array we can type:

```
a = {[1 2] 'hello'; 3  [5;6]};
b = {spiral(3) eye(2) ; 'good' 'bad'};
c = cat(3,a,b);
```

The `cat` function concatenates arrays `a` and `b` along dimension number 3. To visualize this array we can use `celldisp` and `cellplot` as we did before. For example:

`cellplot(c)`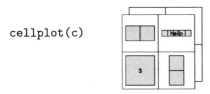

The contents of the cells are only indicated for the front "page" of the multidimensional cell array. To see other pages you can include subscripts into the cell array:

`cellplot(c(:,:,2))`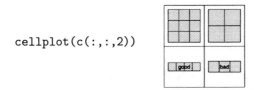

To access cells use curly bracket indexing and to access cell contents use round bracket indexing:

```
>> c{1,2,2}
ans  =
      1    0
      0    1
>> c{1,2,2}(1,:)
ans  =
      1    0
```

29.6   Multidimensional Structures

Multidimensional structures have dot-separated field names, but they are accessed using an arbitrary number of subscripts. For example,

```
>> staff(2,1,2).name = 'Joe Bloggs'
staff =
2x1x2 struct array with fields:
```

```
    name
>> staff(2,1,2).name(5:9)
ans   =
Blogg
```

30   Saving and Loading Data

30.1   MATLAB Format

MATLAB's internal standard for saving and loading data is the mat-file. The **save** command saves all the variables in the workspace to a disk file called `matlab.mat`:

```
>> a = 1;
>> b = 1:10;
>> str = 'hello';
>> save
Saving to: matlab.mat
```

To save in a file with a different name, type:

```
save saved_data
```

To save only some of the variables, add their names:

```
save saved_data a b
```

Mat-files are binary files whose format is described fully in the MATLAB documentation. Such a file is fully portable to MATLAB running on another kind of computer. Information about the kind of computer that MATLAB was running on when it saved the data is stored along with the data. When MATLAB reads in a mat-file, it checks the type of computer that the data were saved on and automatically performs any required manipulations (byte swapping, for example).

To load data from mat-files use the load command:

```
>> clear
>> a = 1;
>> b = 1:10;
>> str = 'hello';
>> save saved_data
>> clear
>> whos
>> load saved_data
```

```
>> whos
  Name        Size          Bytes  Class
   a          1x1               8  double array
   b          1x10             80  double array
  str         1x5              10  char array
Grand total is 16 elements using 98 bytes

>> a
a  =
     1
>> b
b  =
     1   2   3   4   5   6   7   8   9   10
>> str
str  =
hello
```

To save the data as readable text use the -ascii switch:

```
save saved_data_text -ascii
```

In this case the '.mat' extension is not appended to the file name. The ASCII format is best kept for simple cases where your data is in the form of a matrix (or vector or scalar). For example, in this case we have saved the variables a, b, and c in a file that has the following contents:

```
1.0000000e+00
1.0000000e+00    2.0000000e+00    3.0000000e+00    ...
1.0400000e+02    1.0100000e+02    1.0800000e+02    ...
```

The first line is the variable a, the second line is the variable b = [1 2 ... 10], and the third line is the string str = 'hello' converted to its corresponding ASCII values:

```
>> double(str)
ans  =
   104   101   108   108   111
```

If you try to load this data using the load command you will get an error message because the lines have different numbers of values. To load an ASCII file like this, you'll have to write your own loading function using the functions getl etc. described in the next section. If you save an ASCII *matrix*, however, you can load it in again without difficulty:

```
>> clear
>> q = spiral(3)
```

```
q  =
     7      8      9
     6      1      2
     5      4      3
>> save saved_data_text  -ascii
>> clear
>> load saved_data_text
>> whos
  Name                    Size        Bytes  Class
    saved_data_text       3x3            72  double array
Grand total is 9 elements using 72 bytes

>> saved_data_text
saved_data_text  =
     7      8      9
     6      1      2
     5      4      3
```

The data is loaded as a variable with the same name as the file name (no information about variable names are stored in the file).

30.2   Other Formats

You may be presented with some data written by another piece of software that you want to load into MATLAB. In this case you have the following options:

1. You can write a translation program in another language (C or FORTRAN for example) that reads in the data and then writes it to another file that can be read into MATLAB—that is, a mat-file.

2. You can write a MATLAB-callable program (mex-file) that reads in the data and returns appropriate variables in the MATLAB workspace. This is a good option if you already have code to read in the data.

3. You can use one of the functions for reading in standard file formats for images, sounds, spreadsheets,[11] and so on. These are:

   dlmread   Read ASCII data file.
   wk1read   Read spreadsheet (WK1) file.
   imread   Read image from graphics file (JPEG, TIFF, etc.).

---

[11]For Lotus123 spreadsheets you can use the functions wk1read and wk1write. If you use Microsoft Excel, the MathWorks' Excel Link product allows direct communication between Excel and MATLAB. For example, Excel can be used as a front-end for MATLAB; you can call MATLAB functions or graphics routines directly from Excel, or you can access your Excel spreadsheet data directly from MATLAB.

> auread   Read SUN ('.au') sound file.
> wavread Read Microsoft WAVE ('.wav') sound file.
> readsnd Read SND resources and files (Macintosh only).

4. You can write an m-file to read the data, using fopen, fread, and associated functions.

In this section we consider item (4). The functions available are

Category	Function	Description
Open/close	fopen	Open file
	fclose	Close file
Binary I/O	fread	Read binary data from file
	fwrite	Write binary data to file
Formatted I/O	fscanf	Read formatted data from file
	fprintf	Write formatted data to file
	fgetl	Read line from file, discard newline character
	fgets	Read line from file, keep newline character
String Conversion	sprintf	Write formatted data to string
	sscanf	Read string under format control
File Positioning	ferror	Inquire file I/O error status
	feof	Test for end-of-file
	fseek	Set file position indicator
	ftell	Get file position indicator
	frewind	Rewind file
Temporary Files	tempdir	Get temporary directory name
	tempname	Get temporary file name

Following is an example of how some of these functions are used.

Example: fscanf

Suppose we have some data in a file formatted as follows:

```
10/06    11:18:00    -34.855     151.3057    216.4  70.91  -61.23  0.29
10/06    11:18:01    -34.85554   151.30649   214.8  71.38  -60.8   -0.88
10/06    11:18:02    -34.85609   151.30727   212.7  71.86  -60.64  -1.64
10/06    11:18:03    -34.85664   151.30807   210.8  72.4   -60.35  -1.67
10/06    11:18:04    -34.85717   151.30887   209.7  72.83  -60.06  -1.33
```

The data consists of a date string with a slash separator, a time string with colon separators, and then six numbers separated by white space. The function fscanf is used for reading formatted ASCII data such as this from a file. Suppose this file is called asc.dat. First, we must open this file for reading using the fopen command:

```
fid = fopen('asc.dat');
```

The `fopen` command returns a *file identifier*, `fid`, which is an integer that must be used as the first argument of any subsequent file-reading or file-writing function that uses the file `asc.dat`. If the file cannot be opened (for example, if it does not exist or exists in a directory that cannot be found by MATLAB), then a value `fid = -1` is returned. Once the file is opened the data can be read using the following command:

```
>> a = fscanf(fid,'%d/%d %d:%d:%d %g%g%g%g%g%g');
>> size(a)
ans =
    55    1
```

The `fscanf` command has read in all the data up to the end of the file. In the file are 11 numbers per line (2 numbers in the date, plus 3 in the time, plus 6 other numbers), and there are 5 lines, for a total of 55 data values; these have been read into a column-vector called `a`. The format string '%d/%d %d:%d:%d %g%g%g%g%g%g' means "look for two decimal numbers separated by slashes, skip some whitespace, then look for three decimal numbers separated by colons, skip some more whitespace, then look for six general format floating point numbers" (see the section on string conversion on page 79). `fscanf` reads in such numbers until the end of the file, or you can put in a parameter to read in a certain number of values. We only now need to reshape the vector `a` to a matrix having 11 columns

```
N = length(a)/11;
a = reshape(a,11,N)';
```

The date and time values are in the first five columns:

```
>> a(:,1:5)
ans =
    10    6   11   18    0
    10    6   11   18    1
    10    6   11   18    2
    10    6   11   18    3
    10    6   11   18    4
```

And the remaining values are

```
>> a(:,6:11)
ans =
  -34.8550  151.3057  216.4000   70.9100  -61.2300    0.2900
  -34.8555  151.3065  214.8000   71.3800  -60.8000   -0.8800
  -34.8561  151.3073  212.7000   71.8600  -60.6400   -1.6400
  -34.8566  151.3081  210.8000   72.4000  -60.3500   -1.6700
  -34.8572  151.3089  209.7000   72.8300  -60.0600   -1.3300
```

31   Handle Graphics

Handle Graphics is MATLAB's system of creating and manipulating computer graphics. The system is "object oriented", meaning that it is based around a hierarchy of objects that represent various graphical elements. These elements all have a certain state, or "appearance", defined by a list of handle properties, and they can be changed by a number of different methods. The properties of objects can be set at creation or they can be modified afterwards. The complete set of graphical objects is shown in this diagram.

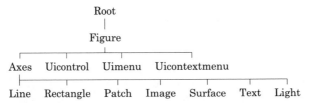

Graphical objects lower down in the hierarchy inherit many of their properties from those higher up. Objects that are immediately below another in the hierarchy are said to be that object's *children*; the object immediately above another is said to be that object's *parent*.

Rich graphics contain many of these elements, with the design enhancing the overall utility of the display. For example, this diagram shows some common Handle Graphics objects. The frame around the

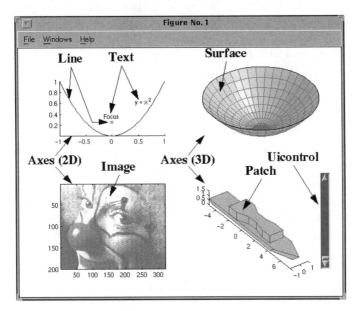

display and the enclosed area is the Figure object; that is, the window on the screen in which the graphics are displayed. Many Figure objects can exist at the same time, and the Figure's handle is the number shown in the window's title bar (usually it is an integer, $1, 2, \ldots$). Above Figure objects in the hierarchy is the Root object, which is the entire computer screen. Only one of these can exist, and its handle is the number zero. In this Figure are four Axes objects (one is invisible): two are three-dimensional and two are two-dimensional. The top left-hand Axes object contains two Text objects ('Focus' and '$y = x^2$'), and two Line objects (the parabola and the single point marked by an '×'). These two Line objects look different because they have different "LineStyle" and "MarkerStyle" properties; more on this later.

31.1   Get and Set

The commands **get** and **set** are used to find out about the state of graphics elements (**get**) and to change those elements' properties (**set**). For example, we will create a simple plot, and use **get** and **set** to change some of the plot's properties. The plot is simply:

```
t = linspace(0,10,50);
plot(t,sin(t))
```

Suppose we want to plot the points themselves as well as the line joining them. We could create a new plot by typing `plot(t,sin(t),'-o')`, but we can do the same thing by first getting the handle of the Line object and setting its Marker property, which is initially `'none'`, to o:

```
h = get(gca,'children');
set(h,'Marker','o')
```

The command **gca** that appears here as an argument to the **get** command is the Get Current Axes command: it returns the handle of the current Axes object, where "current" means the last Axes that were plotted to or clicked on with the mouse. We could have combined the two commands and eliminated the need to actually assign a value for the current Axes' handle:

```
set(get(gca,'children'),'Marker','o')
```

In this case there is only one "child" of the current axes; if there were more, then a vector of handles would be returned and each would have its **Marker** property changed to `'o'`.

There are usually a great many properties associated with a given graphical object. For example the x- and y-axis limits are given by two separate properties, xlim and ylim. Continuing with the example above:

```
>> get(gca,'xlim')
ans  =
      0      10
>> get(gca,'ylim')
ans  =
     -1      1
```

The locations of the x-axis tick marks are another property:

```
>> get(gca,'xtick')
ans =
     0     2     4     6     8     10
```

The width of the line used to draw the axes is

```
>> get(gca,'linewidth')
ans  =
     0.5
```

(Your line width might be different.) There are many more. To get a complete list of the properties of a graphical object, leave out the property argument in a call to get. For example, the properties of the Axes object are

```
>> get(gca)
          AmbientLightColor = [1 1 1]
          Box = on
          CameraPosition = [3.5 0 17.3205]
          CameraPositionMode = auto
          CameraTarget = [3.5 0 0]
          CameraTargetMode = auto
          CameraUpVector = [0 1 0]
          (and so on)
```

And the properties of a Line object are (carrying on from the sine-wave example above):

```
>> get(h)
          Color = [1 0 0]
          EraseMode = normal
          LineStyle = -
          LineWidth = [0.1]
          Marker = o
          MarkerSize = [6]
```

```
MarkerEdgeColor = auto
(and so on)
```

In the example above we used **get** to get the handle of the line after we created it. If you know that you will want to modify an object's properties, you can assign its handle at creation time by using an output variable. Our example then becomes:

```
h = plot(t,sin(t));
set(h,'Marker','o')
```

This technique works for all of the plotting commands, **surf**, **semilogx**, **image**, and so on.

Another way to set object properties is to call a creation function with a list of property/value pairs at the end of the argument list. Each kind of graphical object (except the Root object) can be created by typing a command with the same name as the object. For example, let us create a set of axes suitable for plotting range–depth data:

```
axes('Position',[.1 .5 .8 .08],'TickDir','out',...
     'YDir','reverse','xax','top')
```

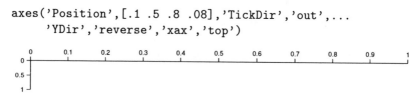

This call to the **axes** function has specified the position property so that we get a short, wide set of axes; the direction of the tick marks so that they stick out of the box instead of into it; the direction of the y-axis is reversed, and the location of the x-axis is on top. The capitalisation of the property names is not important. The name of a property need not be spelled out in full: you can abbreviate it to the shortest unique name. In the above call to **axes**, for example, the **xax** property is the **XAxisLocation**.

If you are unsure of the allowed values for a property, you can get a list of them by typing **set** without actually setting a value. For example, suppose you know there is an Axes property called **XAxisLocation** but you do not know whether to type **'above'**, **'top'**, or **'up'** to get the x-axis drawn at the top. The solution is to type:

```
>> set(gca,'XAxisLocation')
[ top | {bottom} ]
```

The allowed values for the **XAxisLocation** property are **top** and **bottom**. Curly brackets are put around the default setting. If you type **set** without specifying a property you get a list of all the allowed values for every available property:

```
>> set(gca)
        AmbientLightColor
        Box: [ on | {off} ]
        CameraPosition
        CameraPositionMode: [ {auto} | manual ]
        CameraTarget
        CameraTargetMode: [ {auto} | manual ]
        (and so on)
```

The properties for which you can choose values from among a short list of alternatives will be shown. Other properties (for example, the CameraPosition property above) can take on any numerical value, so a list of alternatives is not shown. To get the format of such a property (is the CameraPosition a scalar or a vector?) you can get its current value:

```
>> get(gca,'CameraPosition')
ans =
    0.5000    0.5000    9.1603
```

Some properties are common to all objects. For example, all objects have a "Type" property that specifies what kind of object it is ("Axes", "Figure", "Patch" and so on), a "Parent" property (sometimes empty), a "Visible" property that determines whether you can see it or not, and a "Color" property (fairly obvious). Other properties are unique to a particular kind of object. For example, only line objects have a "LineWidth" property, and only "Figure" objects have an "InvertHard-Copy" property.

Let us now consolidate these ideas with a few examples.

Example: Undo

When building a plot from the command line, it is good to have an "oops" function that gets rid of the last thing you plotted. Let us start by plotting a labelled parabola defined by $f(x) = x^2$.

```
x = -1:.01:1;
f = inline('x.^2');
clf
plt(x,f(x))
```

We use the text command to label the parabola.

`text(-.7,f(-.7),'f(X)')`

But we have made a mistake: the "x" should be lower case. We try to correct it by issuing another `text` command with a lower case "x":

`text(-.7,f(-.7),'f(x)')`

But this has printed over the top of the previous label, making a mess. Without starting again, we can use the `delete` function to delete the text objects, once we know their handles. We can get a list of the handles of the line and the text objects by getting all the children of the current axes:

```
>> h = get(gca,'children')
h  =
    19.0001
    18.0001
    11.0005
```

The variable h is a three-element column vector (the actual values are not important). These are the handles corresponding to the Line object (parabola) and the two Text objects. But which of them is the Line object and which are the Text objects? We can get the object types corresponding to these handles by typing:

```
>> types = get(h,'type')
types =
    'text'
    'text'
    'line'
```

(The variable `types` is returned as a cell array.) A parent's children are always listed in reverse age order: the most recently drawn object appears first—youngest first, oldest last. This tells us that the first two elements of the vector h correspond to the text objects 'f(X)' and 'f(x)', in that order, and the third element corresponds to the parabolic line. We can delete the two text objects by typing:

```
delete(h(1:2))
```

We can now issue the correct `text` command:

text(-.7,f(-.7),'f(x)')

Let us write an m-file to do this automatically. We'll call it oops. When we call oops without any argument, it should delete the last object drawn in the current axes. When called with an integer argument, n, oops should delete the last n objects drawn in the current axes. The following m-file is a possible solution. (The function `nargin` returns the number of arguments with which a function was called.)

```
function oops(N)

% OOPS Delete the last object plotted on the axes.
%       Repeating "oops" erases farther back in time.
%       OOPS does not work for title and labels; to
%       erase these, use "title('')" or "xlabel('')"
if nargin = = 0 N = 1; end
h = get(gca,'children');
delete(h(1:N));
```

Let us see if oops works:

```
clf
plt(x,f(x))
hold on
plt(x,f(x/2))
```

Now we do an oops to get rid of the shallow parabola,

oops

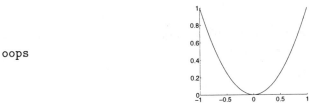

Now plot the shallow one again with a different line style:

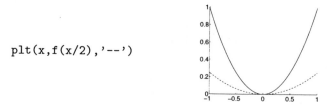

```
plt(x,f(x/2),'--')
```

Let us try calling oops to delete more than one object. (We use a string cell array in the following call to text to label the two curves at once.)

```
x = -1:.01:1;
f = inline('x.^2');
clf
plt(x,f(x),x,sqrt(f(x)))
xt = [-.5 -.5];
yt = [f(-.5) sqrt(f(-.5))];
text(xt,yt,{' |x|' ' x^2'})
```

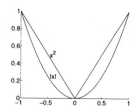

Now get rid of the misplaced labels and try again:

```
oops(2)
text(xt,yt,{' x^2' ' |x|'})
```

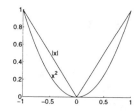

Example: Figure Positioning

In this example we suppose that we want to set up some windows for an application that will have two graphics displays and a graphical user interface. We want the two output displays to occupy the bottom half of the screen and the GUI should occupy the top left-hand corner. We use get and set to control the position of our figures. Let us create a figure and get its "position" property:

```
figure
get(gcf,'position')
ans  =
   291   445   560   420
```

But what do these numbers mean? To find out we need to get the units of measurement of this position:

```
>> get(gcf,'units')
ans  =
pixels
```

Hmm .... What are the available units of measurement?

```
>> set(gcf,'units')
[ inches | centimeters | normalized | points | pixels ]
```

We want to set the units to be normalized: the extremities of the screen will correspond to zero and one.

```
>> set(gcf,'units','norm')
>> get(gcf,'units')
ans   =
normalized
```

The position of the figure in these new normalized units is

```
>> get(gcf,'position')
ans   =
    0.2517    0.4933    0.4861    0.4667
```

This position vector is of the form [left bottom width height]. To create the two figures we first set the position property of this figure to occupy the lower left corner of the screen:

```
set(gcf,'pos',[0 0 .5 .5])
```

We create the other figure and set its position property at the same time:

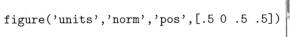

```
figure('units','norm','pos',[.5 0 .5 .5])
```

Exercise 14 *If you look carefully at the two figures you have created in this example, you might notice that the borders of the windows overlap. The reason is that the figure's position property only applies to the area contained within the figure and not to the borders supplied by the computer's windowing system. Assume that these borders are 5 pixels wide on the left, right and bottom edges and 10 pixels wide on the top edge. Write some code to create three figures occupying exactly the bottom-left, bottom-right, and top-right quarters of the screen, with no gap between them or overlap. (The command* `close all` *might come in handy when experimenting with figure creation.) (Answer on page 190.)*

Example: `findobj`

The `findobj` command is used to search through the graphical hierarchy for objects that satisfy particular property values. For example, generate a sphere, a cylinder, and a cone:

```
subplot(131)
sphere
axis equal
ax = axis;
subplot(132)
cylinder
axis equal
axis(ax)
subplot(133)
cylinder([1 0])
axis equal
axis(ax)
```

The three shapes are represented by three surface objects within three axes objects. To get the surface handles by getting the children of the three axes you would need to type three calls to **get**; one for each of the axes:

```
axes_handles = get(gcf,'children');
surf_handle(1) = get(axes_handles(1),'children');
surf_handle(2) = get(axes_handles(2),'children');
surf_handle(3) = get(axes_handles(3),'children');
```

But an easier way to get the surface handles is to use the `findobj` command. Here we use it to find all the objects in the current figure whose **type** property has the value **surface**:

```
surf_handle = findobj(gcf,'type','surface');
```

We can now work with the vector of surface handles to alter all the surfaces at once. Let us make them transparent:

```
set(surf_handle,'FaceColor','none')
```

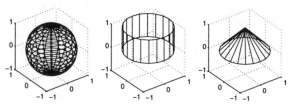

31.2   Default Object Properties

Every graphical object has a set of default property values, so that you are not obliged to spend time thinking about every detail of every graphical object you draw. For example, when you type plot(1:10) you do not necessarily want to think about how thick the line should be, where the tick marks should go, how long the tick marks should be, what colour the axes should be, what MATLAB should do when you click on the line, etc.

When MATLAB creates a graphical object it searches through the successive ancestors (parent, grandparent, etc.) until it either finds a default value defined by the user or a factory-set value. For example, you could define your own default line width for a figure, in which case all new lines drawn in any axes in that figure would have the new line width. Or you could set a default line width in an axes object: such a default would affect lines drawn in that set of axes, but lines drawn in another set of axes would have their widths set by a search through their own, different set of ancestors.

To get a list of factory-defined settings, issue the command get(0,'factory'). You cannot change the factory settings, but you can change the default settings. To get a single factory-defined setting, use the get command, giving it the property name starting with factory, followed by the name of the graphical object (figure, axes, etc.), followed by the name of the property. For example, the factory paper type used for printing figures is

```
>> get(0,'factoryfigurepapertype')
ans   =
usletter
```

Factory settings are not necessarily those that will be used; they may be over-ridden by setting a *default* value. Default values may be changed for your MATLAB installation in either the matlabrc.m file or in your personal startup.m file.

To get a list of default settings, issue the command get(Handle, 'Default'), where Handle is the handle of the object you are interested in. Setting a default value at the Root level (Handle = 0) will affect *all* objects of that type.

To set a default value you use the set command, giving it the handle of the object whose children (grandchildren, etc.) you want affected. You create a default property name by creating a three-part string:

1. Start with the word Default;

2. Add the name of the object you want affected (for example, Line, Surface, Text);

3. Add the property you want to set for this type of object (for example, LineWidth, FaceColor, FontSize).

We fix ideas with an example.

Suppose we do not like the default figure settings that produce plots in a white Axes box on a gray Figure background. Instead you want transparent Axes plotted on a parchment-coloured background. You also want to increase the size of the font used to label the axes. To set these preferences for every graphic you draw, change these default properties:

```
set(0,'DefaultFigureColor',[0.95 0.9 0.8])  % parchment
set(0,'DefaultAxesColor','none')
set(0,'DefaultAxesFontSize',12)
```

(Capitalization is not essential.) The Axes font size setting affects all text associated with Axes objects: tick labels, axis labels and titles, but not Text objects or text on uicontrols. If you want to change these as well, you could issue the commands:

```
set(0,'DefaultUIControlFontSize',12)
set(0,'DefaultTextFontSize',12)
```

These sorts of commands often go in your startup.m file, where they are executed each time MATLAB starts.

31.3   Current Objects

In MATLAB graphics there are alway three "current" objects: the current Figure, the current Axes, and the current Object. The current Figure or Axes objects are the ones that the next gaphical object will be drawn in. The current Object is the last one drawn or clicked on with the mouse. Figures or axes can also be made current by clicking on them with the mouse. We have already used the functions that return the handle of the current Axes object (gca) and the current Figure object (gcf). There is also a gco command that returns the handle of the current Object. These three commands are essentially abbreviations of:

```
gcf:   get(0,'CurrentFigure')
gca:   get(gcf,'CurrentAxes')
gco:   get(gcf,'CurrentObject')
```

Exercise 15 *Can you explain the difference between the following two methods of getting the current figure handle after doing a* close all *? (Answer on page 191.)*

```
>> close all
>> get(0,'currentfigure')
```

```
ans  =
        []
>> gcf
ans  =
        1
```

Exercise 16 *The m-file* `objects.m` *creates a graphic with assorted objects on it. Run* `objects` *and try clicking on the objects to make them current. Investigate what happens when you do* `get(gco,'type')`, *or* `delete(gco)`. *(Answer on page 191.)*

32 Axes Effects

32.1 The Axis Command

Axes objects have many properties you can modify to alter details such as tick mark labels, positioning of axes, direction of tick marks, and so on. These can be changed using the `get` and `set` commands. You can also change the way axes behave using the `axis` command, which is an easy way to set some `axis` and related commands to achieve frequently sought effects. Let us look first at the `axis` command. These are the available options (adapted from the help entry):

`axis([xmin xmax ymin ymax])` sets scaling for the x- and y-axes on the current plot.

`axis([xmin xmax ymin ymax zmin zmax])` sets the scaling for the x-, y-, and z-axes on the current 3-D plot.

`v = axis` returns a row vector containing the scaling for the current plot. If the current view is two-dimensional, v has four components; if it is three-dimensional, v has six components.

`axis auto` returns the axis scaling to its default, automatic mode where, for each dimension, "nice" limits are chosen based on the extents of all line, surface, patch, and image children.

`axis manual` freezes the scaling at the current limits, so that if `hold` is turned on, subsequent plots will use the same limits.

`axis tight` sets the axis limits to the range of the data.

`axis fill` sets the axis limits and `PlotBoxAspectRatio` so that the axis fills the position rectangle. This option only has an effect if `PlotBoxAspectRatioMode` or `DataAspectRatioMode` are manual.

`axis ij` puts MATLAB into its "matrix" axes mode. The coordinate system origin is at the upper left corner. The i axis is vertical and is numbered from top to bottom. The j axis is horizontal and is numbered from left to right.

axis xy puts MATLAB into its default "Cartesian" axes mode. The coordinate system origin is at the lower left corner. The x axis is horizontal and is numbered from left to right. The y axis is vertical and is numbered from bottom to top.

axis equal sets the aspect ratio so that equal tick mark increments on the x-, y-, and z-axis are equal in size. This makes sphere(25) look like a sphere, instead of an ellipsoid.

axis image is the same as axis equal except that the plot box fits tightly around the data.

axis square makes the current axis box a square.

axis normal restores the current axis box to full size and removes any restrictions on the scaling of the units. This undoes the effects of axis square and axis equal.

axis off turns off all axis labeling (including the title), tick marks, and background.

axis on turns axis labeling, tick marks and background back on.

axis vis3d prevents MATLAB from stretching the Axes to fit the size of the Figure window or otherwise altering the proportions of the objects as you change the 3-D viewing angle.

Let us look at some quick examples. Create sine and cosine components and plot a circle:

```
t = linspace(0,2*pi);
x = cos(t);y = sin(t);
plot(x,y)
```

The default behaviour here is such that the data are stretched to fill the rectangular Axes position. To make it look like a circle use:

```
axis equal
```

To get the top half of the circle:

```
axis([-1 1 0 1])
```

Some new data now:

```
t = linspace(0,1);
x = humps(t);
y = humps(t.^2)/2;
plot(x,y)
```

The Axes limits have been set to the next round number in the series of tick marks. To change the scale so that the data fill the whole plot:

`axis tight`

To zoom in on the loop:

`axis([10 25 5 12])`

To zoom back out again:[12]

`axis auto`

Now let us look at some image data:

```
clf
load clown
image(X)
colormap(map)
```

The y-axis here increases from top to bottom: the `ij` axis mode is the default for images. To get the y-axis increasing from bottom to top:

`axis xy`

To go back again:

---

[12]Or you can use the **zoom** function, which initiates a mouse-based zoomer.

`axis ij`

Usually images like this do not need the axes:

`axis off`

A three-dimensional example now:

```
clf
sphere
colormap(fitrange(gray,.5,1))
view(5,5)
axis equal
```

MATLAB has drawn a biggish sphere because the near perpendicular viewpoint allows the axes to fit within the default plotting area. If we change the viewpoint:

`view(45,45)`

the sphere is drawn smaller because the axes are more oblique to the plane. Now go back to the first viewpoint, switch on the `vis3d` axis behaviour, and then return to the second viewpoint:

```
view(5,5)
axis vis3d
view(45,45)
```

The sphere is kept a constant size (cf. plot before last, above), which forces the axes to extend beyond the plotting area (and, in this case, beyond the Figure area too). You should turn on `axis vis3d` whenever you are viewing three-dimensional objects from different angles. In such situations the axes are usually superfluous anyway, so why not get rid of them?

`axis off`

The `axis` command works by changing various properties of Axes objects. If you look inside the `axis`, function (`type axis`) you will see many `set` commands used to change Axes properties. As we said before, the `axis` command gives you an easy way to change frequently used Axes features. Some Axes properties are not part of the `axis` command's functionality; you must change them yourself. For example, when drawing physical objects rather than mathematical abstractions, realism is improved by allowing perspective distortion. Compare these two views of a ship seen from about wharf height:[13]

```
clf
subplot(221);
ship
axis off
pos = [11 2.3 .55];
set(gca,'CameraPosition',pos)
subplot(223);
ship
set(gca,'CameraPosition',pos,...
'Projection','Perspective')
axis off
```

Here `ship` is an m-file on disk that draws the patches representing the ship.

As another example, here is what you might see if you were an ant crawling along a doughnut (the command `torus` is part of the companion software to this book):

```
clf
[x,y,z] = torus(.5,90,1);
surfl(x,y,z,[150,50],[0 1 0 0])
colormap(fitrange(gray,0.5,1))
axis equal
axis off
axis vis3d
pos = [[1 1]*1.1 .7];
set(gca,'CameraPosition',pos)
set(gca,'CameraTarget',[0 .8 .4])
set(gca,'Projection','Perspective')
```

Exercise 17 *When you have driven past a vineyard or an orchard, have you ever noticed the many directions in which the plants seem to line up? Create an evenly spaced grid of points, and see if you can get* MATLAB *to display the same kind of effect. (Answer on page 191.)*

---

[13]The working of `ship.m` is explained in Section 37 on three-dimensional modelling, see page 160.

32.2   Tick Marks and Labels

MATLAB's default behaviour regarding tick marks is to put a reasonable number of ticks at evenly spaced, round number increments that at least span the data. You can change the tick marks using the various tickmark properties:

```
XTick = [1 2 3 4 5]          TickLength = [0.01 0.025]
XTickLabel ['a|b|c|d|e']     TickDir = in
XTickLabelMode = manual      TickDirMode = auto
XTickMode = manual
```

The properties in the first column have equivalents for the y and z axes; the properties in the second column affect the ticks on all axes.

The `TickLength` property must be set to a two-element vector; the first element is the length of tickmarks used for two-dimensional plots and the second element is the length of tickmarks used for three-dimensional plots. The units are normalised to the length of the longest axis:

```
subplot(221)
plt(1:10)
subplot(222)
plt(1:10)
set(gca,'ticklength',[.06 .1])
```

It is more common to want to change to location and labels of the tickmarks. Here are some tickmarks tied to the data:

```
x = sort(rand(1,5));
plt(x)
set(gca,'ytick',x)
axis tight
grid
```

Here is a plot of a sine curve with critical points as tick marks:

```
t = linspace(0,360);
y = sin(t*pi/180);
xt = unique([0:45:360 30:30:360]);
yt = unique(sin(xt*pi/180));
plt(t,y)
axis([0 360 -1 1])
set(gca,'xtick',xt,'ytick',yt,'GridLineStyle','-')
grid
```

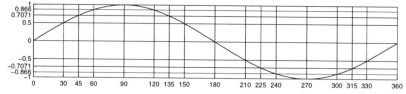

The tick labels can be either numbers or strings. You have a choice of
the following formats:

```
set(gca,'XTickLabel',{'1';'10';'100'}
set(gca,'XTickLabel','1|10|100')
set(gca,'XTickLabel',[1;10;100])
set(gca,'XTickLabel',0:2)
set(gca,'XTickLabel',['1   ';'10 ';'100'])
```

In the second format, the modulus signs "|" separate the tick labels. In
the fifth format you cannot replace the semicolons by commas; if you do
you will be specifying a single tick label equal to the string '1   10 100',
which will be used to label all the tick marks. In another example, here
is how to get months of the year on an x axis:[14]

```
y = [0 3 1 6 5 9];area(y)
str = 'Jan|Feb|Mar|April|May|June';
set(gca,'xtick',1:6,...
'xticklabel',str,...
'xgrid','on','layer','top')
```

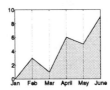

If you specify fewer tick mark labels than there are tick marks, the labels
are recycled:

```
plt(1:10)
set(gca,'ytick',1:10,...
        'yticklabel','a|b|c')
```

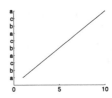

If you want only some tick marks labeled, include blanks (or nothing)
between the modulus signs in the ticklabel setting:

```
plt(1:4)
set(gca,'xtick',1:.2:4,...
'xticklabel','1|||||2|||||3|||||4')
```

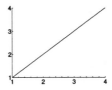

It can sometimes be a good idea to turn off the tick mark labels completely. For example, when stacking plots that cover the same range of
x values:

---

[14]The `datetick` function can also be used in conjunction with the date handling
utility `datenum`.

```
clf
axes('pos',[.2 .1 .7 .4])
x = linspace(0,2);
plt(x,humps(x))
axis tight
zeroaxes
axes('pos',[.2 .5 .7 .4])
plt(x,cumsum(humps(x)))
set(gca,'xticklabel','')
axis tight
```

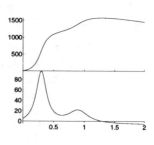

32.3   Subplots

Multivariate data can be displayed by plotting arrays of subplots. For example, a column of x–y plots can represent y plotted as a function of x and z. A sequence of such columns can represent another variable, so that you can see y plotted as a function of x, z, and t. MATLAB's subplot command is an easy way of generating arrays of plots, but you have no control over the precise positioning; the gap between the plots, for example, is not controllable. The m-file pickbox (see companion software) is designed for such cases. You give pickbox the number of rows and columns you want in your array of plots, and pick out the number of the plot you want (in the same way as for subplot). You can also specify the amount of x and y space between the plots as well as the amount of white space around the entire plot array. This space can be used for row and column labels. In the following example we generate samples of the function

$$B(x,y,t) = (1 - e^{-(2x)^2-y^2})/t$$

over a three-dimensional grid of x, y, and t. We display the samples by drawing repeated plots of B as a function of x and arraying them over a matrix of rows and columns indexed by y, and t, respectively:[15]

```
xv = -1:.1:1;
yv = -1:.2:1;
tv = 1:5;
[x,y,t] = ndgrid(xv, yv, tv);
B = (1 - exp(-(2*x).^2 - y.^2))./t;
Nx = length(xv);
Ny = length(yv);
Nt = length(tv);
clf
count = 0;
top = max(max(max(B)));
```

---

[15]The code is given in the companion m-file plotbxyt.

```
for yi = 1:Ny
  for ti = 1:Nt
    count = count + 1;
    pos = pickbox(Ny,Nt,count,0,0,.2);
    ax = axes('pos',pos,...
        'ylim',[0 top],...
        'nextplot','add',...
        'visible','off');
    plt(xv,B(:,yi,ti),'.') % Data
    plt([-1 -1 1],[1 0 0],'k:') % Dotted frame
    if count~ = 51
      set(gca,'xticklabel','',...
          'yticklabel','')
    end
    if count  =  = 51
      axis on
      xlabel('x')
      ylabel('B')
    end
    if count<6
      text(0,1.2,['Time = ' ...
          num2str(tv(ti)) ' s'],...
        'HorizontalAlignment','center')
    end
    if rem(count-1,5) =  = 0
      text(-3,0.5,['y = ' ...
          num2str(yv(yi))])
    end
  end
end
end
```

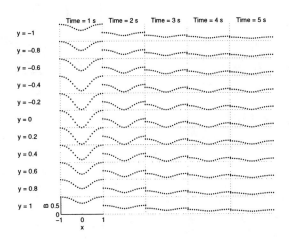

For these kinds of plot arrays it is essential to keep the axes' scales fixed for all the plots. The axes' scales are fixed by setting the YLim property to [0 top] in the call to the axes command (the x scales are the same here). The if statement containing the test count<6 ensures that only the plots in the top row—plot numbers $1, 2, \ldots, 5$—produce the text (created using text commands) on the top row of plots, which indicate the time values for each column. The if statement containing the test rem(count-1,5) = = 0 ensures that only the plots in the left column—plot numbers $1, 6, 11, \ldots, 51$—produce the text indicating the y values for each row.

32.4   Double Axes

To plot more than one set of axes on the same plotting area you can use the plotyy function, which puts separate axes on the left and right of the plot:

```
f = inline('exp(-x.^2)');
g = inline('1 - exp(-x)');
x = linspace(0,1);
plotyy(x,f(x),x,g(x))
```

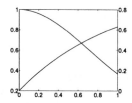

The left hand y axis refers to the function f and the right hand y axis refers to the function g.

Another double-axis technique is to draw axes with two sets of units. The trick here is to create a second set of axes that is very thin:

```
subplot(211)
x = linspace(0,1);
plt(x,humps(x))
xlabel('Range, km')
p = get(gca,'position');
axes('pos',[p(1) .45 p(3) .01],...
'xlim',[0 1]/1.609)
xlabel('Range, miles')
```

32.5   Axes Labels

The various axis-label commands act as expected:

```
plot(1:3)
xlabel('x axis')
ylabel('y axis')
title('Title')
```

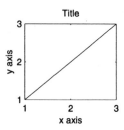

These commands change special Text objects that are associated with the Axes object:

```
>> yl = get(gca,'ylabel');
>> get(yl,'String')
ans  =
y axis
>> set(yl,'Rotation',0)
>> pos = get(yl,'pos');
>> set(yl,'pos',[0.35 3])
```

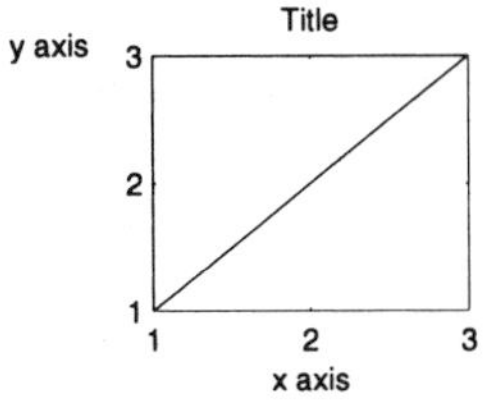

To be friendly to the viewers of your graphs, you should always place your y labels horizontally. Multi-line labels can be done easily using cell arrays:

```
>> str = {'The answer is below:';
['It is ' num2str(pi)]}
str =
    'The answer is below:'
    'It is 3.1416'
>> title(str)
```

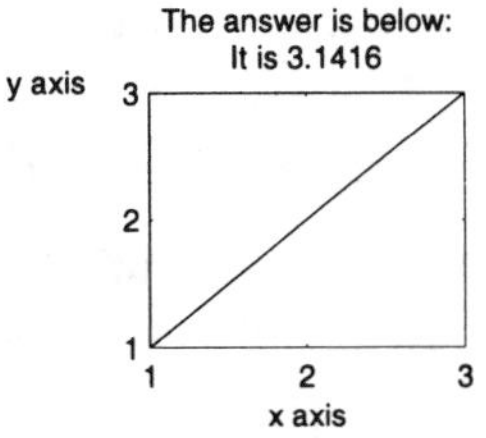

33   Text in Graphics

The commands xlabel, ylabel, zlabel, and title are used to put text in special places around the plot. A general way to place text is to use text commands:

```
x = 0:.01:2;
plt(x,humps(x))
axis tight
[ym,i] = max(humps(x));
str = ['Maximum value: ' ...
              num2str(ym)];
text(x(i),ym,str)
```

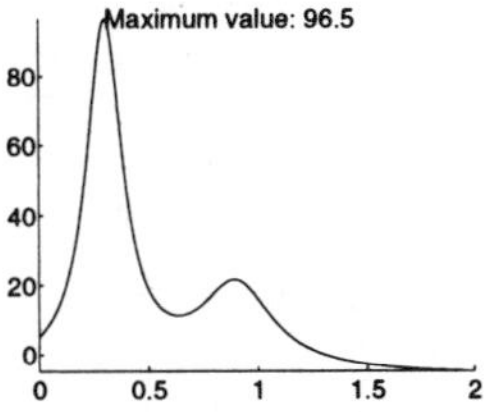

The first two inputs to text define the x and y coordinates of the text's reference point. (The gtext command allows you to define the reference point with the mouse.) You can give a third, z-value, to the text command to place text on three-dimensional plots. Issuing text commands creates Text objects, which have a great many properties (type get(h), where h is the handle of a Text object). Often you want to change the way the text is aligned to its reference point. By default, text is horizontally aligned such that the left-hand edge is near the reference point, and vertically aligned such that the middle of the text is near the reference point. The following diagrams show the effect of changing a Text object's `HorizontalAlignment` and `VerticalAlignment` properties to the indicated values:

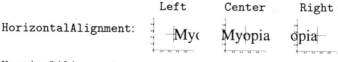

VerticalAlignment:

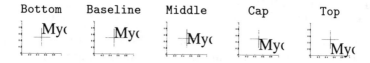

These text objects all have the reference point at (0.5,0.5), indicated by the cross on each little plot.

33.1 Symbols and Greek Letters

The best way to put symbols and Greek letters in your text is to use MATLAB's implementation of the TEX (or LATEX) syntax. TEX is a computer typesetting system for producing high-quality mathematical material. (This book was produced using TEX.) In TEX you produce symbols and Greek letters by typing a backslash "\" followed by the name of the letter or symbol:

Δ	\Delta	$\circ$	\circ	κ	\kappa	ρ	\rho
Γ	\Gamma	♣	\clubsuit	λ	\lambda	$\rightarrow$	\rightarrow
$\Im$	\Im	$\cong$	\cong	$\leftarrow$	\leftarrow	σ	\sigma
Λ	\Lambda	$\cup$	\cup	$\leftrightarrow$	\leftrightarrow	$\sim$	\sim
Ω	\Omega	δ	\delta	$\leq$	\leq	♠	\spadesuit
Φ	\Phi	$\diamondsuit$	\diamondsuit	μ	\mu	$\subset$	\subset
Π	\Pi	$\div$	\div	$\neq$	\neq	$\subseteq$	\subseteq
Ψ	\Psi	$\downarrow$	\downarrow	$\ni$	\ni	$\supset$	\supset
$\Re$	\Re	ϵ	\epsilon	ν	\nu	$\supseteq$	\supseteq
Σ	\Sigma	$\equiv$	\equiv	$\o$	\o	τ	\tau
Θ	\Theta	η	\eta	ω	\omega	θ	\theta
Υ	\Upsilon	$\exists$	\exists	$\oplus$	\oplus	$\uparrow$	\uparrow
Ξ	\Xi	$\forall$	\forall	$\oslash$	\oslash	υ	\upsilon
$\aleph$	\aleph	γ	\gamma	$\otimes$	\otimes	ς	\varsigma
α	\alpha	$\geq$	\geq	∂	\partial	ϑ	\vartheta
$\approx$	\approx	$\heartsuit$	\heartsuit	ϕ	\phi	$\wp$	\wp
β	\beta	$\in$	\in	π	\pi	ξ	\xi
$\bullet$	\bullet	∞	\infty	$\pm$	\pm	ζ	\zeta
$\cap$	\cap	$\int$	\int	$\propto$	\propto		
χ	\chi	ι	\iota	ψ	\psi		

Here are some examples:

```
x = linspace(0,2);
y = 1./(1-x);
plt(x,y)
text(1,75,...
' y \rightarrow \pm\infty')
```

The TEX syntax of '_' and '^' to produce subscripts and superscripts is also supported:

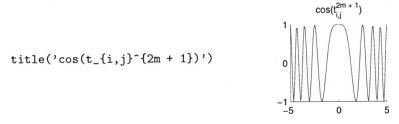

```
t1 = linspace(-5,5);
y = cos(t1.^2);
plt(t1,y)
title('cos(t_1^2)')
```

If the sub- or superscript is more than one character, use curly brackets to define the scope:

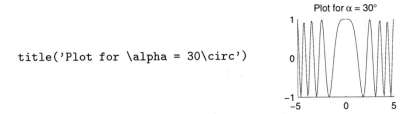

```
title('cos(t_{i,j}^{2m + 1})')
```

For degree symbols (e.g. 30°), use \circ. (In TEX you would use \circ in a superscript, 30°, but doing that in MATLAB makes the degree symbol too high and too small.)

```
title('Plot for \alpha = 30\circ')
```

33.2   Symbols in Tick Labels

To put TEX symbols in tick mark labels you cannot use the commands of the xlabel family; they currently do not interpret the TEX syntax. However, you can replace the default ticklabels with Text objects that contain

the required symbols. The m-file `ticklabelx` does such a replacement.
You might like to look at it in your editor to see how it works. To use
it, you supply a list of tick marks with the TEX symbols included:

```
x = linspace(0,pi,99);
plt(x,sin(x))
xt = 0:pi/6:pi;
set(gca,'xtick',xt)
axis tight
tikstr = {'0','\pi/6','\pi/3',...
'\pi/2','2\pi/3','5\pi/6','\pi'}
ticklabelx(tikstr)
```

33.3   Global Object Placement

Complex displays might contain many Axes or other objects and you may
want to place text, lines, or other objects globally without reference to
any particular Axes object in the display. These objects are children of
Axes objects, so they must be placed relative to some Axes object, but
we can use a trick. The trick is to create an invisible axes object that
covers the entire display and place the required objects inside that. For
example, consider the following technique to create a global title to a
series of subplots:

```
subplot(221),subplot(222)
subplot(223),subplot(224)
axes
str = 'Here are four subplots';
text(.5,1.05,str,...
've','bo','ho','c')
axis off
```

Here, the last four inputs to the `text` command are abbreviations for
the title alignment: `'VerticalAlignment'`, `'bottom'`, `'Horizontal-
Alignment'`, `'center'`.

> **Exercise 18** *Why not use* `title` *instead of* `text` *in the previous
> example? (Answer on page 192.)*

Another way to put text on a graphic independently of the data plotted
is to use normalized position units. The reference point of the text will
then refer to the area occupied by the axes, independently of the data
plotted. Suppose, for example, that you want to print a parameter on a
plot whose axes scaling might change.

```
x = linspace(0,1,200);
y = humps(x);
subplot(221)
plt(x,y)
text(.5,1,'Scale = 1',...
    'units','normalized')
subplot(222)
plt(x/7,y/7)
text(.5,1,'Scale = 1/7',...
    'units','normalized')
```

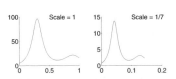

Notice that the `text` commands here use the same positional references, being $x = 0.5$ and $y = 1$ in normalized units.

Another example shows a plot and a zoomed portion:

```
load clown
subplot(221)
imagesc(X)
colormap(map)
axis image off
hold on
plot([150 230 230 150 150],...
     [100 100 60  60 100])
subplot(223)
imagesc(X(60:100,150:230))
axis image off
axes
axis manual
hold on
x = [.195 0    NaN .301 .419];
y = [.793 .348 NaN .793 .348];
plt(x,y)
axis off
```

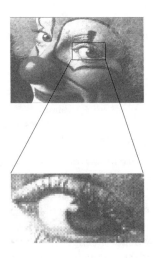

The statement `axis manual` is needed here to freeze the axes limits at their default values. To produce the x and y data for the zoom-lines, I used the `ginput` command to obtain the coordinates with the mouse. The `ginput` command gets input from the current axes. So if you want to add more points to the invisible axis, you must make it visible again, otherwise your mouse click will be interpreted with reference to the last plotted (visible) axes.

34 Graphical User Interfaces

A graphical user interface (GUI) is a system of graphical elements that allow a user to interact with software using mouse operations. There are three ways to make graphical user interfaces:

1. Use graphical elements that serve no other purpose than to allow the user to interact (virtual buttons, switches, knobs, sliders, pup-up menus, and so on). (See `demo` for examples of these.)

2. Use graphical elements that perform a dual function: display data and interaction. For example, a plotted line can both display data and can alter data when a user clicks on the line and drags it to a new position. (See `sigdemo1` in the signal processing toolbox for an example.)

3. Use mouse-downs, drags, and mouse-ups anywhere within the Figure to perform an action. (For example, typing `rotate3d` whenever you are displaying a three-dimensional plot allows you to click and drag to change the viewpoint.)

The first group of GUI elements (buttons, etc.) are the easiest to work with, so we deal with those first.

There are three GUI-specific graphical objects in MATLAB: uicontrols, uimenus, and uicontextmenus.

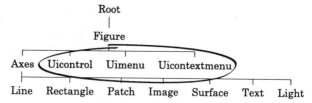

These are at the same level as Axes objects in MATLAB's object hierarchy. They are children of Figures; they have no children. Their appearance and behaviour are defined by their property values. We first look at the different styles of uicontrol. Then we will look at how you can program a uicontrol to do something by setting its `callback` property. Finally, we will go through the various uicontrols in a bit more detail, before considering uimenus. Uicontextmenus control MATLAB's behaviour when you do a "right-click" (or equivalent menu-getting click on your system) on a graphical object. They will not be described here, but you can find a description in the helpdesk entry under Handle Graphics Objects.

If you type `uicontrol`, you will get MATLAB's default uicontrol (we assign its handle to h for later use):

h = uicontrol;

As usual, `get(h)` will give you a list of properties for this object. An important property for uicontrols is the *style* property. The style of this object is

```
>> get(h,'style')
ans  =
pushbutton
```

(Try pushing the button.) The available styles are

```
>> set(h,'style')
[ {pushbutton} | togglebutton | radiobutton | checkbox
  | edit | text | slider | frame | listbox | popupmenu ]
```

The following table shows the possible uicontrol styles:

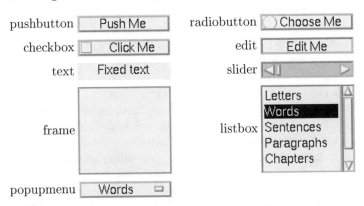

(The togglebutton looks like a normal pushbutton but it stays pushed until you click it again.) The appearance of these uicontrols depends on the windowing system of your computer, but their behaviour in MATLAB from one kind of computer to another is always the same.

34.1   Callbacks

You specify what happens when a uicontrol is activated by setting its CallBack property. Callbacks are statements that get executed in the MATLAB workspace (the command window) when a user interface element is activated. As a simple example consider:

```
uicontrol('String','Do plot','CallBack','plot(humps)')
```

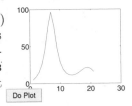

This creates a pushbutton uicontrol (the default) containing the text "Do plot". When this button is pressed with the mouse, the command plot(humps) is executed in the MATLAB workspace. Try it now and you should see a plot of the humps function appear.

   The callback string can be any MATLAB expression or function call. The following simple GUI creates three buttons to create a plot of sin(x), cos(x), and tan(x). The buttons call the MATLAB function ezplot with the appropriate trigonometric function as an input. The double quotes '' produce a single quote in the callback string (see the section on strings, page 74).

```
uicontrol('Callback','ezplot(''sin(x)'')', ...
   'Position',[508 351 51 26],'String','Sine');
uicontrol('Callback','ezplot(''cos(x)'')', ...
   'Position',[508 322 51 26],'String','Cos');
uicontrol('Callback','ezplot(''tan(x)'')', ...
   'Position',[508 293 51 26],'String','Tan');
```

This plot shows the Figure as it appears when you press the 'Tan' button. For simple GUIs the direct definition of callbacks used above is sufficient, but for more complex actions you generally want to execute an m-file as a callback. You can execute a separate m-file for each button in your

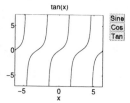

GUI, but this leads to a great many separate m-files associated with a single GUI. A better technique is to use *switchyard programming*.
In switchyard programming you send all your callbacks to a single m-file and change the input to the m-file, depending on which button was pressed. The m-file contains all the code for all the buttons; the appropriate code for a given button is selected by a logic test within the m-file. We adapt the trig-function plotting GUI above to this technique. The m-file is as follows:

```
function trigplt(action)
if nargin = = 0 % Create the GUI:
  uicontrol('Callback','trigplt Sine',...
              'Position',[508 351 51 26],'String','Sine');
  uicontrol('Callback','trigplt Cosine',...
              'Position',[508 322 51 26],'String','Cos');
  uicontrol('Callback','trigplt Tangent',...
              'Position',[508 293 51 26],'String','Tan');
else % Perform the action:
  x = linspace(0,2*pi);
  switch(action)
    case 'Sine'
      y = sin(x); titstr = 'y = sin(x)';
    case 'Cosine'
      y = cos(x); titstr = 'y = cos(x)';
    case 'Tangent'
      y = tan(x); titstr = 'y = tan(x)';
  end
  plot(x,y)
end
```

This m-file is given in the companion software file `trigplt.m`. If you type `trigplt`, the m-file will execute the part that creates the GUI, since it was called with no input arguments (`nargin = 0`). Pressing the buttons will calculate the appropriate trig function and produce the plot.

Callbacks are fastest when they are implemented as function calls; do not implement your callbacks as script m-files or as an `eval` of a string. The reason is that MATLAB compiles a function the first time it is encountered, whereas m-files and `eval`s are interpreted line by line.

The button-style uicontrols (pushbuttons, radiobuttons, and checkboxes) are used by simply clicking on them with the mouse. Others need more interaction: you must choose an item from a list (listboxes or popupmenus) or specify a numeric value (sliders) or type in text (edit boxes). We discuss briefly the operation of each of these different kinds of uicontrols.

34.2   UIControls

Edit Boxes

Edit boxes are designed to read in a piece of typed text. The text inside an Edit box is accessed via the box's String property:

```
h = uicontrol('style','edit','String','Hello');
```
Hello

You can change it using the `set` command:

```
set(h,'string','Bye')
```
Bye

or you can click in the box and change it by typing something else. You can access what has been typed into an edit box by getting its string property. After typing `qwe` into the box you can type:

qwe
```
>> get(h,'string')
ans =
qwe
```

Numbers typed into edit boxes remain strings until you convert them to numbers:

10.3
```
>> x = get(h,'string')
x =
10.3
>> x+1
ans =
    50    49    47    52
>> str2num(x) + 1
ans =
   11.3000
```

Text

Good GUIs have instructive text that indicates the function of a uicontrol. These can be placed with the text-style uicontrol. In the following GUI the "Name:", "Address:", and "Sex:" labels are three separate uicontrols of Text style.

```
uicontrol('Pos',[110 280 60  19],'Style','text','String','Name:');
uicontrol('Pos',[175 280 246 19],'Style','edit');
uicontrol('Pos',[110 262 60  19],'Style','text',...
          'String','Address:');
uicontrol('Pos',[175 262 246 19],'Style','edit');
uicontrol('Pos',[110 243 60  19],'Style','text','String','Sex:');
uicontrol('Pos',[175 243 121 19],'Style','radiobutton',...
          'String','Male');
uicontrol('Pos',[301 243 121 19],'Style','radiobutton',...
          'String','Female');
```

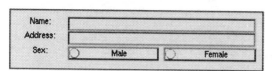

Frame

The frame style uicontrol is not an object that is meant to be interacted
with. It is a decoration. For example, in the GUI in the previous section
the text labels stand apart from the uicontrols they refer to because the
figure color is not the same as the background colour. You can get a
more integrated appearance by adding a frame:

```
uicontrol('Position',[99   231 332 77],'Style','frame');
uicontrol('Position',[110 280 60  19],'Style','text',...
          'String','Name:');
uicontrol('Position',[175 280 246 19],'Style','edit');
uicontrol('Position',[110 262 60  19],'Style','text',...
          'String','Address:');
uicontrol('Position',[175 262 246 19],'Style','edit');
uicontrol('Position',[110 243 60  19],'Style','text',...
          'String','Sex:');
uicontrol('Position',[175 243 121 19],'Style','radiobutton',...
          'String','Male');
uicontrol('Position',[301 243 121 19],'Style','radiobutton',...
          'String','Female');
```

You must issue the commands to draw the uicontrols over the frame
after you issue the frame command, otherwise the frame will obscure
the uicontrols. Some people like to divide their GUIs into sections using
frames. In this GUI are three frames:

The two inner frames are labeled by placing text style uicontrols with the strings `Personal Details` and `Tax Details` at appropriate positions. Too many frames clutter the GUI. It is often better to use space between groups of controls to divide them into logical groups.

Slider

Sliders are designed to allow input of a value between two limits. You input a value by clicking on the central bar and dragging it; by clicking anywhere between the central bar and the end of the slider, resulting in a big jump; or by clicking on the arrows at the ends of the slider, resulting in a little jump. Fiddle with the following slider to see how it behaves:

```
h = uicontrol('style','slider');
set(h,'pos',[50 200 450 40])
set(h,'callback','disp(get(h,''value''))')
```

This slider has its `callback` property set so as to display the value of the slider in the command window. The callback is executed when (1) you release the central bar, (2) you click on the blank area, or (3) you click on the arrow.

You may want to display the value of a slider in an edit box, or to use the value typed into an edit box to alter the position of the slider. To do this you have to get an edit box and a slider to talk to each other to reveal their properties. The edit box must have a callback that tells the slider its `string` property, and the slider must have a callback that tells the edit box its `value` property. The following piece of code achieves this effect:

```
clf
hsl = uicontrol('Position',[200 260 200 20], ...
       'Style','slider','Value',0.5,...
       'CallBack',...
       'set(hed,''String'',num2str(get(hsl,''value''),2))');
hed = uicontrol('BackgroundColor',[1 1 1], ...
       'Position',[200 240 70 20], ...
```

```
'String','0.5','Style','edit',...
'CallBack',...
'set(hsl,''Value'',str2num(get(hed,''String'')))');
```

The value from the edit box is transmitted to the slider when you press
return, tab, click outside the edit box, or move the mouse outside the
GUI's window. The slider's callback must convert the slider's value—
a number—to a string before setting the edit box's string property. In
principle the converse is not true. That is, the slider's value property can
be passed directly to the edit box's string property and the number will
be displayed automatically as a string. But we have included a numerical
conversion using num2str(...,2) to limit the number of decimal places
of the displayed value to 2. (What happens when you type nonnumeric
text into the edit box?)

The default limits of a slider are set to a minimum of zero and a
maximum of one. These can be changed via the slider's min and max
properties: the slider's **value** property is scaled proportionally.

Listbox

Listboxes let you choose from among a scrollable list of alternatives. The
list of alternatives is set by the listbox's **string** property. The options
for the format of the string are the same as for the axis tick labels (see
page 125); you can specify the alternatives in any of the following ways:

```
set(h,'String',{'Red';'Green';'Blue'})
set(h,'String','Red|Green|Blue')
set(h,'String',[1;10;100])
set(h,'String',1:3)
set(h,'String',['Red  ';'Green';'Blue '])
```

For example, to bring up a list of colour options:

```
h = uicontrol('pos',[168 219 89 116], ...
    'Style','listbox', ...
    'String',{'Red';'Green';'Blue'});
```

If the list is too wide or high for the listbox, MATLAB adds scroll sliders:

```
set(h,'string',{'Red';'Green';'Blue';...
    'Pale Goldenrod';'Orange';'yellow'})
```

The item selected within the listbox is accessed via its "value" property. For example, after selecting a colour from the list you can extract it using the following commands:

```
>> str = get(h,'string');
>> str(get(h,'value'))
ans =
     'Pale Goldenrod'
```

Popup menu

Popup menus are similar to listboxes in that they allow you to choose from among a list of alternatives, but only one item is shown at a time; the others become visible only when you press the button. Assuming your listbox is still present from the previous example, you can convert it to a popup menu by typing the following:

```
set(h,'style','popup',...
    'pos',[168 219 145 32])
```

We changed the size (position) at the same time to make it look more like a standard button. The user's choice is accessed by the popup menu's "value" property, as for listboxes.

34.3 Exclusive Radio Buttons

Radio buttons can be used to offer a choice of one, and only one, item from among alternatives. Think of a car radio with buttons for the different radio channels: when you press one button in, the corresponding channel only is selected. MATLAB's radio buttons do not automatically behave this way. You may want to allow more than one radio button to be pressed at a time. But if you do want exclusive radio buttons you must implement them with appropriate callbacks. One way to do it is as follows:

```
function exradio(action)
if nargin = = 0
  clf
  uicontrol('Position',[200 321 90 25], ...
      'String','JJJ', ...
      'Style','radiobutton',...
      'CallBack','exradio(1)')
  uicontrol('Position',[200 296 90 25], ...
      'String','ABC-FM', ...
      'Style','radiobutton',...
      'CallBack','exradio(1)')
  uicontrol('Position',[200 271 90 25], ...
      'String','SAFM', ...
      'Style','radiobutton',...
      'CallBack','exradio(1)')
  uicontrol('Position',[200 246 90 25], ...
      'String','5AD', ...
      'Style','radiobutton',...
      'CallBack','exradio(1)')
else
  h = findobj('style','radiobutton');
  ind = find(h~ = gco);
  set(h(ind),'value',0)
end
```

Calling the function exradio with no arguments draws the GUI and sets up the callbacks. The callbacks are identical for all the buttons: they simply call exradio with a single input argument. When any button is clicked, the else code is executed: it first finds all the radiobuttons and returns their handles in the vector h. Then a vector, ind, of the elements of h that are not equal to the Current Object (got by the call gco) is defined. The radio button that has just been clicked on will be the Current Object. If a radio button is pushed, its "value" toggles between zero and one. The set command then sets the "value" property of all the other radio buttons to zero.

34.4   Variables in GUIs

Globals

Variables in the MATLAB workspace are not visible to function m-files. If you use a function m-file to implement a GUI, you often need to access variables that won't be visible to the function unless you make them so. To explain this, consider the example of the exclusive radio buttons given in the last section. Suppose we want to get rid of the findobj command in the else section of the code by putting the radio buttons'

handles into a vector when they are defined. That is, we want to modify
the code as follows:

```
function exradio2(action)
if nargin = = 0
  clf
  h1 = uicontrol('Position',[200 321 90 25], ...
      'String','JJJ', ...
      'Style','radiobutton',...
      'CallBack','exradio2(1)');
  :
  :
  h4 = uicontrol('Position',[200 246 90 25], ...
      'String','5AD', ...
      'Style','radiobutton',...
      'CallBack','exradio2(1)');
  % Save radio button handles in h for later use.
  h = [h1 h2 h3 h4];
else
  ind = find(h~ = gco);
  set(h(ind),'value',0);
end
```

This implementation will not work because function variables are local to
the function and do not persist from one function call to another. When
a radio button is pushed it will issue a callback to `exradio2`, which will
go to the `else` section of code where it will crash because the variable
h will not be defined. One way to implement the idea correctly is to
declare the vector h to be *global*. Global variables are visible to all other
functions that declare them global, and thus they will be visible between
one function call and the next. The correct implementation will be

```
function exradio2(action)
global h
if nargin = = 0
  clf
  :
  :
```

You can even access such global variables from the MATLAB workspace
(the command window) by declaring them global there.

Variables in UserData

A problem with global variables is that they are vulnerable to being
cleared by the user from the workspace. If the user clears the global
variables that a function expects to be present, then the function will

fail. One place to put variables that need to be visible to functions and where they cannot be cleared is in graphical objects' UserData property. Every graphical object has a UserData property, which can be accessed with the get and set commands. The UserData property can be used to store any MATLAB variable.

34.5   The Tag Property

As we saw above in the radiobutton example, it is useful to have some means of finding the handle to an object without explicitly saving it as a variable. The Tag property lets you uniquely name an object. You can then find the object anywhere in your code by looking for the object with that name. For example, we could find the handle of the button called 'JJJ' using Handle = findobj('tag','JJJ');. (Of course, you must set the tag property beforehand.)

34.6   UIMenus

By default, MATLAB's Figures come with a menu at the top. The menu items are File, Windows, and Help.

figure

You can add your own items to this menu or you can delete it and put your own in its place. To delete the default menu, you need to set the Figure's menubar property to 'none' (set it to 'figure' to bring it back again):

set(gcf,'menubar','none')

To add your own menu use the uimenu command. The text that appears on the menu is set by the menu's label property; what happens when you select the menu item is set by the menu's callback property. Menus can be children of Figures or of other menus; in the latter case you get submenus, or "walking" menus. The following example produces a menu of options to change the colour of the Figure.

```
f = uimenu('Label','Figure Colour');
    uimenu(f,'Label','Default',...
        'Callback','set(gcf,''color'',''default'')',...
        'Accelerator', 'D');
```

```
uimenu(f,'Label','Black',...
    'Callback','set(gcf,''color'',''k'')');
uimenu(f,'Label','White',...
    'Callback','set(gcf,''color'',''w'')');
uimenu(f,'Label','Gray',...
    'Callback','set(gcf,''color'',[.5 .5 .5])');
g = uimenu(f,'Label','Other...','Separator','on');
uimenu(g,'Label','Parchment',...
    'Callback','set(gcf,''color'',[.95 .9 .8])');
uimenu(g,'Label','Vellum',...
    'Callback','set(gcf,''color'',[.9 .9 .8])');
uimenu(g,'Label','Cream',...
    'Callback','set(gcf,''color'',[.95 .9 .75])');
```

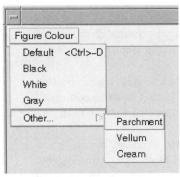

The uimenu item that sets the Figure colour to the default has its "accelerator" property set to "D", meaning that when control-D is pressed while the cursor is in the Figure window, the callback will be executed; in this case the Figure will go to its default colour. The "Other ... " menu item has its "separator" set to "on", which draws the line above its label. A clear figure (clf) command will clear user-created uimenus like this one. The command colormenu is worth looking at as another example of a simple uimenu.

Exercise 19 *Generate the following menu:*

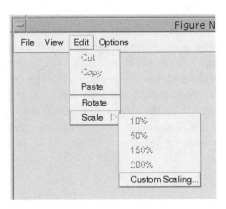

(Answer on page 192.)

34.7   Fast Drawing

For fast drawing of graphics (e.g., when animating), consider the following extract from the Mathworks' web site:

> Draw movable or changing objects with the `EraseMode` property set to `xor` or `background`. This prevents re-rendering the axes when changing these objects. `EraseMode` is a property of all objects that are children of axes (line, text, surface, image, patch).
>
> Set `DrawMode` (an axes property) to `fast`. This prevents MATLAB from sorting three-dimensional objects, which can speed things up significantly. The side-effect is that three-dimensional surface plots will not be drawn properly in this mode.
>
> Set `BackingStore` (a figure property) to `off`. This should give roughly a factor of two speed-up in normal drawing but turns off the instantaneous update that normally occurs when windows are uncovered.
>
> Set `NextPlot` (a figure property) to `new` when creating a GUI. That way when you make plots from the command window they do not appear in the axes of the GUI's figure window.
>
> Wherever possible, recycle figure windows by using the `visible` property of figures instead of creating and destroying them. When done with the window, set `visible` to `off`; when you need the window again, make any changes to the window and then set `visible` to `on`. Creating figure windows involves much more overhead than setting their `visible` property to `on`.

There is also a property of Figures called `doublebuffer`. Setting this property to `'on'` can reduce the flicker when redrawing a figure.

When redrawing a figure from within a loop, MATLAB will wait until the final run through the loop before rendering the graphic. If you want to see intermediate results you need to force MATLAB to dedraw the graphic at that point. Use the `drawnow` command to do this forcing. In the following example, animation is used to demonstrate the sampling problem known as aliasing.[16]

```
clf
set(gcf,'doublebuffer','on')
h = plot(0,0,'.');
for i = 1:1000
  t = linspace(0,2*pi,i);
```

---

[16]This example due to C. Moler, `comp.soft-sys.matlab` newsgroup.

```
  set(h,'xdata',t,'ydata',sin(370*t))
  drawnow
end
```

The example should update the data without flicker. If you then set the
`doublebuffer` property to `off`, you should see the window flicker as the
new data updates the plot.

34.8 Guide

When writing code to generate GUIs you can very quickly get bogged
down in messy coding. For example, it is tedious to calculate the "position" properties of the various elements so that they line up nicely. The
task of generating GUIs is made much easier with the MATLAB tool
`guide`. The name `guide` is short for Graphical User Interface Devel-
opment Environment. `Guide` is a GUI for drawing GUIs. With it you
select uicontrols, position them with the mouse, align them with the
alignment tool, and set the properties with a graphical property editor.
I used `guide` to produce most of the GUI examples so far. When you
type `guide` at the MATLAB prompt, the current figure shows a grid in
the background, showing that it is now a `guide`-controlled figure, and
the `guide` control panel appears. Try typing `guide` now and you should
see the following control panel on your screen:

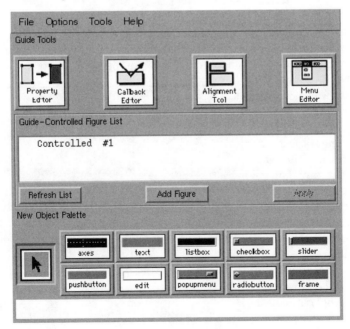

The four big buttons at the top of the control panel bring up four other
control panels. The listbox in the middle displays which figures are being

controlled by `guide` and which are active (meaning that the uicontrols are no longer moveable by guide). The buttons at the bottom of the control panel allow you to select different uicontrols and then draw them on the figure with the mouse. Clicking on the button with the picture of a cursor arrow on it allows you to select objects in the controlled figure.

Exercise 20 *As with any good GUI, reading about what it does is not as effective as actually using it. Play with* `guide` *now and see if you can make a spectacular GUI. This one might inspire you:*

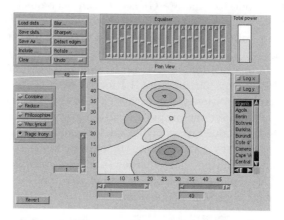

The property editor allows you to select an object, then view and change its properties. If you type the first few unique characters into the property box and press return, the property name will be automatically completed and the value will be shown in the edit box. You can select more than one object, in which case the values that any particular property has in common for all the selected objects will be shown; values that are different will be indicated by question marks. For example, this property editor has five radio buttons

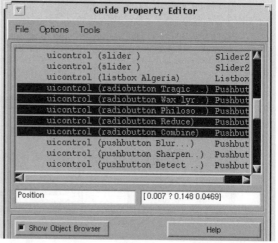

selected, and their "position" property is being shown. The position coordinates are the same except for the y position because all the buttons are the same size and aligned at the same x position. If you edit this property, all of the selected items will be affected.

Guide can be used for *any* MATLAB Figure, not just one that has uicontrols or uimenus. For example, you can use guide to move around a set of axes on a figure.

34.9 Other Aids

A suite of tools to help you program GUIs can be found in the uitools subdirectory. Type help uitools to get a list of these tools and a short summary of what they do. For example, the btngroup function can be used to automatically create a group of buttons in which only one button is allowed to be down at a time. Another possibility is a "flash" button that presses in and immediately pops back out again. Button appearance can also be customised using the btngroup function.

When building a GUI, prototypes can be saved using print -dmfile. Elements in your GUI that you have produced from the command window will then be saved in the m-file.

For further information on GUIs, see the manual *Building GUIs with MATLAB*.

35   Printing Graphics

When you type print at the MATLAB prompt, the current figure is printed on your default printer. The plot is printed so that the aspect ratio matches that seen on the screen for the default settings, and it is placed centrally on the page oriented like a portrait: the long dimension of the page upright:

```
plot(humps)
print
```

If your MATLAB has been installed properly, you should find the plot centred for the paper size you are using. If it is not, put the following statement into your startup.m file:

```
set(0,'DefaultFigurePaperType','a4letter')
```

or substitute your correct paper size (see below for a list of options).

There are many ways you can control how a graphic is printed. The following is a list of figure properties that have to do with printing and their options:

```
InvertHardcopy: [ {on} | off ]
PaperUnits: [ {inches} | centimeters
            | normalized | points ]
PaperOrientation: [ {portrait} | landscape ]
PaperPosition
PaperPositionMode: [ auto | {manual} ]
PaperType: [ {usletter} | uslegal | a3 | a4letter
            | a5 | b4 | tabloid ]
```

We won't look at these in detail to start with; instead we'll look at an easy way to change these properties to get commonly wanted effects.

35.1   Print Size: `Orient`

The `orient` command is an easy way of setting the various figure properties to get rudimentary control of printed output. There are three kinds of orientation, each illustrated below.

Tall

```
plot(humps)
orient tall
print
```

The `tall` orientation can be used when you have plots stacked on top of each other:

```
for i = 1:40,subplot(10,4,i),end
orient tall
print
```

Landscape

```
plot(humps)
orient landscape
print
```

Portrait

The default orientation is "portrait", which can be restored using:

```
plot(humps)
orient portrait
print
```

35.2 Print Size: WYSIWYG

By default, the size of the printed figure does not depend on the size of the figure on the screen. If you stretched the window so that the figure looked like this on the screen, [figure], the printed page would still look like this: [figure]. This is because the figure is scaled when printing to occupy a rectangular area whose size and position are defined by the figure's `paperposition` property. The factory default value for this property is:

```
>> get(gcf,'paperunits')
ans  =
inches
>> get(gcf,'paperposition')
ans  =
    0.25    2.5    8.0    6.0
```

The `paperposition` vector, like an `axes` position vector within a figure, has the form [`left bottom width height`]. The `left` and `bottom` values are taken relative to the lower left-hand corner of the page. The figure window's border is not considered part of the figure for the purposes of printing. To make MATLAB automatically calculate the printed figure's position so that it is the same size as the figure window on the screen (excluding the window border), set the figure's `PaperPositionMode` property to `auto`:

```
set(gcf,'paperpositionmode','auto')
```

Now the figure that looks like this on the screen [figure] will look like this when printed [figure]. With the `paperpositionmode` set to auto, you must make sure that the figure's size on the screen is not too big to fit on the printed page.

35.3 Including Figures in Other Applications

General Comments

The best quality printed figures are produced using PostScript printers. To include a postscript file in another document, you should print from MATLAB using one of the Encapsulated PostScript formats (colour, level 1, or level 2). Use the `-deps` option when issuing a `print` command.

Another option is to use an image format output such as a bitmap, JPEG, or TIFF file. These will not give high quality curves (you will see the pixels), but they are fine for images. Bitmap files can be produced using the `-dbitmap` on Microsoft Windows, or by using a screen grab utility on other platforms (for example, snapshot or xv on the UNIX machines, or Snapz on the Macintosh).

When incorporating large images into other documents, consider using bitmaps instead of PostScript files. For large images, bitmap files are much smaller and may enable you to get around memory problems when printing large files.

In some cases large z-data images produced using MATLAB's `image` function are better rendered using the `contourf` (filled contour) function. The final graphic will take longer to calculate in MATLAB, but if you print it in PostScript, the file will not only be much smaller, but the quality will be higher because you won't see the pixelated contour edges.

Finally, consider the viewers of your graphics, and how they will view them. If your graphics will be included in text that will end up as a report, article, book, etc., your graphics should be the best you can make them. Include plenty of rich detail in your graphics, with user-friendly text put at appropriate places on the display. Simple graphics, such as line plots, can be shrunk to quite a small size (somewhere between postage stamp and postcard size) without loss of detail. Such shrinking will enable you to put more graphics on a text page, or more explanatory text. Try to put your graphic on the same page (or double page spread) as the text that discusses it. Your readers won't be obliged to flip pages or, worse, search through all the graphics collected as afterthoughts at the end of the document.

If your graphic forms part of a personal presentation (the dreaded overhead projector), a different set of considerations apply. Your graphics should be big enough to be seen from the back of the room (is the text big enough, are the lines thick enough?). You will be there to personally explain the graphic's features and significance, but such an explanation will be transient and linear; your audience won't be free to look at the graphic at their own pace, or go back to it later on.

PostScript and Encapsulated PostScript

As mentioned above, the highest quality results will be achieved using PostScript output, and printed on a PostScript printer. PostScript files are text files containing page layout commands in Adobe's PostScript language. Encapsulated PostScript (EPS) files are best for including in other documents; they are single page PostScript files that include information about how big the graphic is. If you print a graphic using MATLAB's plain PostScript option (`print -dps file`), the first few lines of *file*.ps will look like this:

```
%!PS-Adobe-2.0
%%Creator: MATLAB, The Mathworks, Inc.
%%Title: file.ps
%%DocumentNeededFonts: Helvetica
%%DocumentProcessColors: Cyan Magenta Yellow Black
%%Pages: (atend)
%%BoundingBox: (atend)
%%EndComments

%%BeginProlog

% MathWorks dictionary
/MathWorks 150 dict begin
```

The file begins with the characters %!PS, which, when sent to a printer, tells the printer to interpret the rest of the file as PostScript language commands, and not as text to be printed. Lines beginning with percent characters "%" are comments and are, except for the first line, ignored by the printer. Actual PostScript commands begin with forward slashes "/". The line here that reads:

```
%%BoundingBox: (atend)
```

says that the bounding box is to be found at the end of the file.

If you print the same graphic using MATLAB's Encapsulated PostScript option (print -deps *file*), the first few lines of *file*.eps will look like this:

```
%!PS-Adobe-2.0 EPSF-1.2
%%Creator: MATLAB, The Mathworks, Inc.
%%Title: file.eps
%%DocumentNeededFonts: Helvetica
%%DocumentProcessColors: Cyan Magenta Yellow Black
%%Pages: 1
%%BoundingBox:    74    210    549    589
%%EndComments

%%BeginProlog

% MathWorks dictionary
/MathWorks 150 dict begin
```

The crucial difference is that the bounding box information appears near the start of the file. The bounding box is of the form xll yll xur yur, where:

xll is the x coordinate of the lower left corner of the graphic,
yll is the y coordinate of the lower left corner of the graphic,
xur is the x coordinate of the upper right corner of the graphic,

yur is the y coordinate of the upper right corner of the graphic. The units are in points (there are 72 points per inch and 2.54 centimetres per inch). For example a bounding box specification of:

`%%BoundingBox:        100 100 172 172`

would refer to a graphic occupying a one-inch square that is 100 points from the bottom left-hand corner of the page.

Software to include Encapsulated PostScript graphics uses the bounding box to correctly position the graphic on the screen.

LaTeX: Version 2e

The inclusion of Encapsulated PostScript files in LaTeX2e documents is fully described in *Using EPS Graphics in LaTeX2e Documents*, by Keith Reckdahl, available via FTP as `epslatex.ps` from `ftp://ftp.tex.ac.uk/tex-archive/info/` and from other sites of the Comprehensive TeX Archive Network (CTAN) on the Internet. Reckdahl's article gives a very thorough description of importing EPS graphics, and associated LaTeX commands. The standard technique is to use the `graphicx` package, which implements the `\includegraphics`' command and options. (A description of the `graphicx` package can be found in *Packages in the "graphics" bundle*, by David Carlisle, available as `grfguide.ps` or `grfguide.tex` from `ftp://ftp.tex.ac.uk/tex-archive/macros/latex/packages/graphics/` and from other CTAN sites. The following is a summary of these two articles.

To include a MATLAB Encapsulated PostScript file (or any other standard Encapsulated PostScript file) in LaTeX2e you can use the commands:

```
\documentclass{article}
\usepackage{graphicx}
\begin{document}
   \includegraphics{file.eps}
\end{document}
```

The graphic will be included at its natural size. The `.eps` extension can be left out of the file specification, and full path names are allowed. Usually you do not want the graphic to appear at its natural size; you will want to scale it so that its width is fixed at some value and its height is scaled proportionally. To do this, use commands such as these:

`\includegraphics[width = 4cm]{file}` (width is 4 cm)
`\includegraphics[width = \textwidth]{file}` (width is the same as the text)
`\includegraphics[width = 0.5\textwidth]{file}` (width is half the width of the text)

`\includegraphics[width = \textwidth-4cm]{file}` (width is 4 cm less than the width of the text (needs the `calc` package))

Other optional arguments to the `\includegraphics` command allow you to specify the height of the graphic, or the total height (height plus depth), to scale relative to the graphic's natural size, to rotate, clip, trim, and shift the graphic, and to get many other effects. For example, in this book I put the output and the command(s) that produced it side by side using two minipage environments:

```
\begin{flushleft}
  \begin{minipage}{30mm}
    \begin{verbatim}
plt(1:10)
    \end{verbatim}
  \end{minipage}
  \begin{minipage}{0.2\textwidth}
    \includegraphics[width = \textwidth]{onetoten}
  \end{minipage}
\end{flushleft}
```

to produce:

`plt(1:10)`

Inside a `minipage` `\textwidth` is the width of the `minipage` which, in this case, is a fifth of the document's `\textwidth`.

LaTeX: Version 2.09

To include a MATLAB figure in a LaTeX2.09 document you can use the `epsf` package. Print the figure using the `-deps` option in MATLAB's `print` command. This will create an Encapsulated PostScript file in the current directory with the name, say, `graphic.eps`. Put `\usepackage{epsf}` after your `\documentclass` declaration at the top of your input file. Figures can then be included using commands such as:

`\epsfxsize = 0.3\textwidth]{graphic.eps}`

Many LaTeX users like to put their graphics in floating figure environments, with captions and a centred graphic. This is how to do it:

```
\begin{figure}
  \begin{center}
    \leavevmode
    \epsfxsize = 0.5\textwidth\epsffile{graphic.eps}
```

```
  \end{center}
  \caption{This is the figure's caption.}
  \label{graph}
\end{figure}
```

The command \epsfxsize = 0.5\textwidth makes the graphic's width equal to half the value of \textwidth. The y size of the graphic will be scaled proportionally. The epsf function uses the bounding box comment in the .eps file to determine the size of the graphic.

The final (PostScript) output can be previewed on the screen using previewing software such as PageView (Sun), Ghostview, or GNU Ghostscript (multiplatform freeware).

Microsoft Word

Word 7 To include a MATLAB graphic in Microsoft Word, print an Encapsulated PostScript file from MATLAB. Then, from within Word, create a frame by selecting "Insert"→"Frame" and drag the mouse so that the frame is as big as you want the graphic. Then insert the Encapsulated PostScript file into the frame by choosing "Insert"→"Picture ... ". Select the appropriate file using the dialog box that appears. The graphic will not be viewable on the screen in the Word document but it will appear when printed.

Word 97 In Word 97, choose "Insert"→"Picture"→"From File ... " and select the .eps file from the dialog box. Once the picture is inserted you can resize it by clicking on it and dragging the nodes that appear at the corners and edges.

Microsoft PowerPoint

In PowerPoint, select "Insert"→"picture ... " and select the file you want to insert. Encapsulated PostScript and bitmap files generally produce good results. A PostScript figure will not appear on the screen, except as a box outline, but a bitmap will appear. However, a PostScript figure will be higher quality when printed. If you want to overlay explanatory text, arrows, etc. from within PowerPoint, use the bitmap format.

CorelDraw

CorelDraw is able to import files of many different formats. From the File menu choose "Import ... " and select the file. You may need to select a file type from the "Files of type" popup menu. The Encapsulated PostScript format will not appear on the screen, except as a box outline, but it will be high quality when printed. Bitmaps do appear on the screen, but are lower quality when printed.

Bitmap Import On PCs choose MATLAB's -dmeta option when using the print command to produce a file with the extension .emf. Then from CorelDraw select "File"→"Import ... " to get a file finder dialog box. Select the .emf file you created with MATLAB and choose the "Windows Metafile (WMF)" option in the "Files of Type" box. Click the "Import" button and CorelDraw will return you to your document where you must drag out a rectangle to define the size of the imported graphic. Once you drag the rectangle, the MATLAB graphic will appear in bitmap form.

Vector Graphic Import Another option with CorelDraw is to import an HPGL file printed using the -dhpgl option from MATLAB. This is a format designed for Hewlett-Packard pen plotters, and has the advantage that the imported image can be edited from within CorelDraw. Follow the steps as for bitmaps above, but use the "HPGL Plotter File (PLT)" option in the "Files of Type" box. You have the option here of changing the pen colours used by CorelDraw's emulation of an HP pen plotter. This format works well for graphics that could easily be plotted on a pen plotter. Line drawings work well but images do not. Surface plots are initially imported with no hidden line removal, but if you change the fill colour to white with CorelDraw, hidden lines will be removed. To do this fill colour change, ungroup the graphic, select the surface, click the right mouse button, and choose "Properties", then choose white as the fill colour. If you have problems importing with this format, try opening the .hgl file in a text editor and deleting the last line.

36 Irregular Grids

In this section we discuss how to deal with data that is defined over an irregular grid. In Section 7.6 on page 32 we saw how do handle nonrectangular grids, but these were still regular in the sense that the x and y data grids could be defined using rectangular matrices (though the resulting geometrical domains did not have to be rectangles).
Some irregular three-dimensional data is supplied with MATLAB in the data file seamount.mat. Load the data and plot the points:

```
>> load seamount
>> whos
  Name       Size     ...
  caption    1x229    ...
```

```
   x              294x1      . . .
   y              294x1      . . .
   z              294x1      . . .
Grand total is 1111 elements
>> plot3(x,y,z,'.')
>> axis tight
```

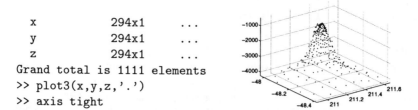

Here we have 294 measurements of x (latitude), y (longitude), and z (height above sea level, which is negative). They represent measurements of a mountain under the sea.[17] The data are stored as column vectors of x, y, and z values. Suppose we want to plot a surface and a contour map representing this seamount data. If you try to type surf(x,y,z) or contour(x,y,z) with this data, you will not get any meaningful plot. Two ways of generating the desired plots are as follows:

1. Interpolate the data over a rectangular grid.

2. Use triangular gridding instead of rectangular gridding.

Let us look at each of these.

36.1   Interpolation over a Rectangular Grid

We continue the example above and define vectors of uniformly spaced points between the minimum and maximum values of x and y:

```
xiv = linspace(min(x),max(x),50);
yiv = linspace(min(y),max(y),50);
```

Each of these vectors has 50 elements. We now use the griddata interpolation function to do two things: (1) create matrices of the x and y grids that correspond to a rectangular grid over the vectors xiv and yiv, and (2) interpolate the data over this new rectangular grid. In the call to griddata that follows, we need to transpose the vector yiv because griddata expects it to be a column vector in this case:

```
>> [xi,yi,zi] = griddata(x,y,z,xiv,yiv');
>> whos
   Name          Size          Bytes  Class
   caption       1x229           458  char array
```

_____

[17]The reference can be found by typing the caption variable:
```
>> caption
caption =
Parker, R. L., Shure, L. & Hildebrand, J., "The application of inverse
theory to seamount magnetism", Reviews of Geophysics vol 25, pp 17-40,
1987.  x is latitude (degrees), y is longitude (degrees), z is negative
depth (meters).
```

x	294x1	2352	double array
xi	50x50	20000	double array
xiv	1x50	400	double array
y	294x1	2352	double array
yi	50x50	20000	double array
yiv	1x50	400	double array
z	294x1	2352	double array
zi	50x50	20000	double array

Grand total is 8711 elements using 68314 bytes

We now have three new matrix variables xi, yi, and zi that correspond to the interpolated data. We make a plot of the original data and the interpolated surface:

```
plot3(x,y,z,'o')
hold on
surf(xi,yi,zi)
colormap(autumn)
axis tight
```

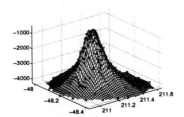

Where the points of the rectangular interpolation grid lie outside the convex hull defined by the data, the values are interpolated as NaN and are omitted from the surface plot. There are a variety of ways to do the interpolation; these are described in the help entry for griddata. We can use the interpolated data to plot a contour map of the seamount:

```
contour(xi,yi,zi)
```

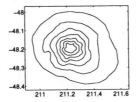

36.2 Triangular Gridding

MATLAB comes equipped with the following functions for use in defining triangular grids:

```
griddata  delaunay  trimesh  dsearch
convhull  voronoi   trisurf  tsearch
```

The idea is that for any set of points (distinct and with no colinear subsets) in two dimensions, a set of triangles can be defined such that (1) no points lie within any triangle's circumcircle and (2) the set completely covers the convex hull of the points. This idea is illustrated in this diagram. Such a triangular grid can be calculated for our seamount example of the last section. The triangles are defined using the `delaunay` function:

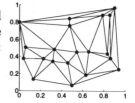

```
>> tri = delaunay(x,y);
>> tri(563:end,:)
ans  =
     2     4     7
     1     2     6
     8     3    11
    25   245    59
```

We have displayed the last few lines of the $M \times 3$ matrix `tri`, which defines the triangles by a set of triplets that are indices into the x and y vectors. For example, the four triangles we have displayed in `ans` are

$$1: \quad \langle (x(2), y(2)) \quad (x(4), y(4)) \quad (x(7), y(7)) \rangle$$
$$2: \quad \langle (x(1), y(1)) \quad (x(2), y(2)) \quad (x(6), y(6)) \rangle$$
$$3: \quad \langle (x(8), y(8)) \quad (x(3), y(3)) \quad (x(11), y(11)) \rangle$$
$$4: \quad \langle (x(25), y(25)) \quad (x(245), y(245)) \quad (x(59), y(59)) \rangle$$

We can use this triangulation matrix to plot a surface of the seamount data; each face of the surface is one of the triangles:

```
trisurf(tri,x,y,z)
hold on
plot3(x,y,z,'o')
axis tight
```

The functions `trisurf` and `trimesh` do not create surface objects; rather, they create `patch` objects.

37   Three-dimensional Modelling

37.1   Patches

In this section we discuss the representation of real-world objects. Such objects are built up using their faces (the six faces of a cube, for example). In MATLAB "faces" are patches, and are defined using the `patch` command. Patches are blobs of coloured light (or ink) that are defined by vertex points. The line between the vertices is the patch's *edge* and the enclosed area is the patch's *face*. Before talking about three-dimensional objects we discuss the simpler two-dimensional patch.

Simple Two-Dimensional Patches

To define a simple patch, specify the x and y coordinates and the face colour:

```
x = [0 1 1 0];
y = [0 0 1 1];
patch(x,y,'y')
axis([-2 2 -2 2])
```

The colour of the edge is black by default. The `patch` function automatically closes the edge if the last vertex is not the same as the first vertex. Patches are usually defined with noncrossing boundaries, but boundaries can cross if required:

```
x(4) = 2; y(4) = .5;
clf
patch(x,y,'y')
```

(The `patch` command is a low-level command, which means that it generates a patch in the current axes without first clearing the axes. That is the reason for the `clf` above.)

Between patches whose boundaries cross and those whose boundaries do not cross are the patches whose boundaries "touch". These can be used to create patches with holes:

```
x = [0 .5 .5 .4 .5 .6 .5 .5 1 .5];
y = [0  0 .1 .2 .3 .2 .1  0 0  1];
clf
h = patch(x,y,'y');
```

The tell-tale line between the outer boundary and the hole can be deleted by making it either invisible or the same colour as the face:

```
set(h,'edgecolor','none')
```

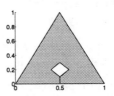

(The command `set(h,'edgecolor','y')` achieves the same effect.) But it would be nice to leave the edge boundary drawn around the shading; you just have to plot a line with the right points:

```
xt = x([1 9 10 1]);
yt = y([1 9 10 1]);
xh = x(3:7);
yh = y(3:7);
hold on
plot(xt,yt,'k',xh,yh,'k')
```

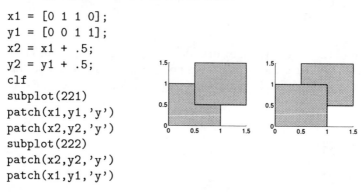

Overlapping patches are drawn in order:

```
x1 = [0 1 1 0];
y1 = [0 0 1 1];
x2 = x1 + .5;
y2 = y1 + .5;
clf
subplot(221)
patch(x1,y1,'y')
patch(x2,y2,'y')
subplot(222)
patch(x2,y2,'y')
patch(x1,y1,'y')
```

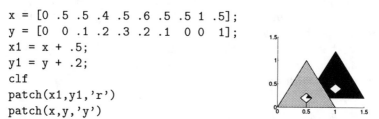

When patches with holes overlap, the one underneath shows through the hole:

```
x = [0 .5 .5 .4 .5 .6 .5 .5 1 .5];
y = [0  0 .1 .2 .3 .2 .1  0 0  1];
x1 = x + .5;
y1 = y + .2;
clf
patch(x1,y1,'r')
patch(x,y,'y')
```

Patches defined by vectors that contain NaNs leave a gap in the edge at the NaN point and leave the enclosed region unfilled:

```
t = linspace(0,2*pi,10);
x = cos(t);y = sin(t);
subplot(221)
patch(x,y,'y')
x(5) = NaN;y(5) = NaN;
subplot(222)
patch(x,y,'y')
```

Three-dimensional Patches

Three-dimensional patches are produced by giving the patch command x, y, and z data. The following generates an inclined triangle:

```
xt = [0 1 .5];
yt = [0 0  1];
zt = [0 0  1];
clf
patch(xt,yt,zt,'y')
view(3)
box
xyz
```

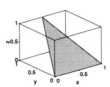

Remember that `patch` is a low-level graphics function, so we must set the view to three dimensional by hand.

A plane is defined by three points, but four points need not lie in a plane. In such a case the patch may look a bit strange, depending on the viewing angle:

```
x = [0 1 1 0];
y = [0 0 1 1];
z = [0 0 0 1];
clf
subplot(221)
patch(x,y,z,'y')
view(-40,10);box;xyz
subplot(222)
patch(x,y,z,'y')
view(33,30);box;xyz
```

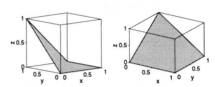

Three-dimensional patches should be planar. The above case, for example, is better done as two patches:

```
x1 = [0 1 1];y1 = [0 0 1];z1 = [0 0 0];
x2 = [0 1 0];y2 = [0 1 1];z2 = [0 0 1];
clf
subplot(221)
patch(x1,y1,z1,'y')
patch(x2,y2,z2,'y')
view(-40,10);box;xyz
subplot(222)
patch(x1,y1,z1,'y')
patch(x2,y2,z2,'y')
view(33,30);box;xyz
```

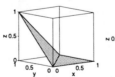

Complex three-dimensional objects should be built up using non-intersecting three-dimensional patches. These can be drawn with a single call to the `patch` function, in which x, y, and z are matrices. Each column of the matrix will define a face. For example, consider the triangular pyramid:

```
x = [0   0  .5  0
    .5  .5  1   1
     1  .5  .5  .5];
y = [0   0   1   0
     1   1   0   0
     0  .5  .5  .5];
z = [0   0   0   0
     0   0   0   0
     0   1   1   1];
clf
h = patch(x,y,z,'y')
view(3);box;xyz
```

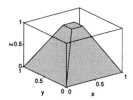

Exercise 21 *Define* x*,* y*, and* z *matrices to draw a truncated square pyramid (answer on page 192):*

```
patch(x,y,z,'y')
view(3);box;xyz
```

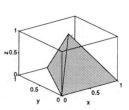

Using x, y, and z matrices to draw objects results in the same vertex being listed as many times as the number of faces that share the vertex. A more compact way of drawing such multifaceted patches is to define a matrix of vertices and a matrix of faces.

Consider again the above triangular pyramid and which is shown here with labelled corners. The vertices are numbered from 1 to 4 and the faces can be defined by specifying the order of joining the vertices. For example, the base is formed by joining the vertices **1**, **2**, and **3**, and

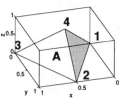

the white front face "**A**" is formed by joining the vertices **2**, **3**, and **4**. The vertices and faces can be defined by the following matrices:

$$\text{Vertices} = \begin{array}{ccc} x & y & z \\ \begin{pmatrix} 0 & 0 & 0 \\ 0.5 & 1 & 0 \\ 1 & 0 & 0 \\ 0.5 & 0.5 & 1 \end{pmatrix} & \begin{array}{l} \leftarrow \text{vertex } \mathbf{1} \\ \leftarrow \text{vertex } \mathbf{2} \\ \leftarrow \text{vertex } \mathbf{3} \\ \leftarrow \text{vertex } \mathbf{4} \end{array} \end{array}$$

$$\text{Faces} = \begin{pmatrix} 1 & 2 & 3 \\ 1 & 2 & 4 \\ 2 & 3 & 4 \\ 1 & 3 & 4 \end{pmatrix} \begin{array}{l} \leftarrow \text{base} \\ \\ \leftarrow \text{face } \mathbf{A} \end{array}$$

The MATLAB code to draw the triangular pyramid using these matrices is

```
vertices_mx = [ 0   0   0
                .5   1   0
                 1   0   0
                .5  .5   1];
faces_mx = [1   2   3 % base
            1   2   4
            2   3   4 % Face 'A'
            1   3   4];
clf
patch('Vertices',vertices_mx,...
'faces',faces_mx,'FaceColor','y')
view(162,44)
box;xyz
```

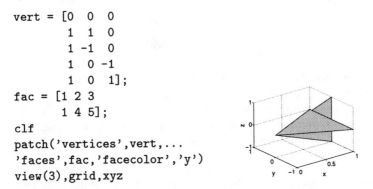

When drawing three dimensional objects, beware of intersecting patches. Each patch is drawn in its entirety, so intersecting patches often look strange. For example, here are two triangles that intersect along their symmetry axes:

```
vert = [0   0   0
        1   1   0
        1  -1   0
        1   0  -1
        1   0   1];
fac = [1 2 3
       1 4 5];
clf
patch('vertices',vert,...
'faces',fac,'facecolor','y')
view(3),grid,xyz
```

Explore this graphic by typing rotate3d and moving the viewpoint with the mouse. You should see that you never get a realistic image. A better way to create the required display is to generate four nonintersecting triangles:

```
vert2 = [0   0   0
         1   1   0
         1  -1   0
         1   0  -1
         1   0   1
         1   0   0];
```

```
fac2 = [1 2 6
        1 6 3
        1 4 6
        1 6 5];
```

The resulting display is now rendered correctly no matter what the viewing angle:

```
clf
patch('vertices',vert2,...
'faces',fac2,'facecolor','y')
view(3),grid,xyz
```
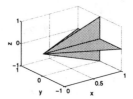

Patch Colouring

Simple solid colouring of patches can be specified using the named colours, as we did for the yellow ('y') patches of the previous section. You can also use arbitrary RGB colours. Here is a patch that should be orange on your display:

```
xt = [0 1 .5];
yt = [0 0 1];
clf
h = patch(xt,yt,[1 .4 0])
```

Patches have a number of properties that control how they are coloured. By using coloured patches, you can make pictures of objects that are colour-coded to some quantity you want to display. For example, the stress of a bent bar could be presented as the colour of the bar. Here is a list of patch colour properties and a description of what they do (taken from the *Using MATLAB Graphics* manual).

CData Specify single, per face, or per vertex colours in conjunction with x, y, and z data.

CDataMapping Specifies whether colour data is scaled or used directly as indices into the Figure colormap.

FaceVertexCData Specify single, per face, or per vertex colours in conjunction with faces and vertices data.

EdgeColor Edges can be invisible, a single colour, a flat colour determined by vertex colours, or interpolated colours determined by vertex colours.

FaceColor Faces can be invisible, a single colour, a flat colour determined by vertex colours, or interpolated colours determined by vertex colours.

`MarkerEdgeColor` The colour of the marker, or the edge colour for filled
   markers.

`MarkerFaceColor` The fill colour for markers that are closed shapes.

The key to patch colouring is to define a colour matrix of the right size
for the type of colouring you want to apply. The following tables illustrate the kinds of patch colouring possible with the `FaceVertexCDdata`
property of patches. The matrices are shown assuming that the patch
has N_f faces and N_v vertices.

Indexed Colours:

	Single colour	One colour per face	One colour per vertex
	C	$\begin{bmatrix} C_1 \\ C_2 \\ \vdots \\ C_{N_f} \end{bmatrix}$	$\begin{bmatrix} C_1 \\ C_2 \\ \vdots \\ C_{N_v} \end{bmatrix}$

In the table above, the numbers C are indices into the current colour
map and in the following table R, G, and B represent red, green, and
blue intensity values between 0 and 1:

True (RGB) Colours:

Single colour	One colour per face	One colour per vertex
$\begin{bmatrix} R & G & B \end{bmatrix}$	$\begin{bmatrix} R_1 & G_1 & B_1 \\ R_2 & G_2 & B_2 \\ & \vdots & \\ R_{N_f} & G_{N_f} & B_{N_f} \end{bmatrix}$	$\begin{bmatrix} R_1 & G_1 & B_1 \\ R_2 & G_2 & B_2 \\ & \vdots & \\ R_{N_v} & G_{N_v} & B_{N_v} \end{bmatrix}$

We now give some examples of colouring effects (for a detailed description of the patch colouring properties see the *Using MATLAB Graphics*
manual).

Example: Stressed Cable   Suppose that you are an engineer working
on a problem involving a cable under stress. You want to display the
shape of the cable and colour the cable according to the stress at each
point along it. We implement the display using a `patch`. First we
generate some x, y, and z data to define the shape of the cable. For
illustrative purposes let us assume the cable shape is a one-turn helix:

```
t = linspace(0,2*pi,20);
x = cos(t);
y = t;
z = sin(t);
```

We will use the vertex-and-faces method of specifying the patch. The vertex and face matrices are

```
v = [x' y' z'];
f = 1:20;
```

We need a colormap of length 20 to colour our 20 data points:

```
fvc = summer(20);
```

We generate a patch whose vertices are coloured according to the matrix fvc:

```
clf
h = patch('Vertices',v,...
'Faces',f,...
'FaceVertexCdata',fvc,...
'FaceColor','flat',...
'EdgeColor','flat',...
'Marker','o',...
'MarkerFaceColor','flat');
view(44,18),axis equal,box
```

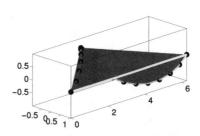

The patch looks a bit strange because its edge is a helix and not a planar shape. The face of this patch is the same colour as the first vertex. If we wanted the patch to be shaded the same way as its edge, we could set its `facecolor` to `interp`:

```
set(h,'FaceColor','interp')
```

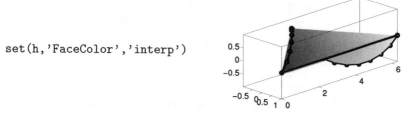

But in this case our data is contained within the patch's edge so we can set the `facecolor` to `none`:

```
set(h,'FaceColor','none')
```

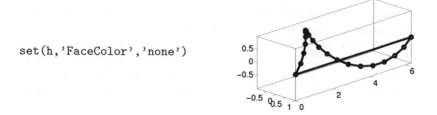

The `patch` function has automatically joined the first and last points, which we do not want to do. We can get rid of that line by setting the final colour of the `FaceVertexCData` matrix to be `NaN`:

```
>> fvc(20,:) = NaN
fvc =     0    0.5000    0.4000
       0.0526    0.5263    0.4000
       [. . .]
       0.8947    0.9474    0.4000
       0.9474    0.9737    0.4000
          NaN       NaN       NaN
>> set(h,'facevertexcdata',fvc)
```

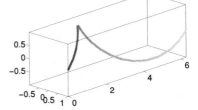

As for most other instances of plotting NaNs, MATLAB handles not-a-number elements by leaving them out. Our final plot of the cable omits the points:

```
>> set(h,'marker','none')
```

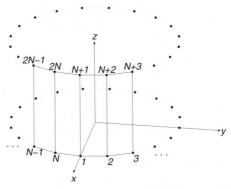

Example: Coloured Cylinder Suppose a cylindrical section of pipe is heated and that it develops a temperature distribution such that the temperature anywhere on its surface depends on the distance from the point of heating. We will use a single call to the patch command to draw the cylinder and display the temperature coded as different colours on the cylinder's surface. We define the cylinder by defining the two rings at its ends. We will use the vertex-and-faces method of spec-

ifying the patch, and number the vertices according to the scheme shown in this diagram. The vertices at the bottom are numbered from 1 to N; the vertices at the top are numbered from $N + 1$ to $2N$. The first face will be formed by joining vertices 1, $N + 1$, $N + 2$ and 2. The second face will be formed by joining vertices 2, $N+2$, $N+3$ and 3; and so on. The final

face will be formed by joining vertices N, $2N$, $N + 1$ and 1. N is equal to 20 in the diagram shown here. We start by defining the x, y, and z coordinates that we need:

```
N = 20;
dt = 2*pi/N;
t = 0:dt:(N-1)*dt;
x = [cos(t) cos(t)];
y = [sin(t) sin(t)];
z = [zeros(size(t)) ones(size(t))];
```

The matrix of vertices is

```
vert = [x' y' z'];
```

The matrix of faces must be defined so that each row gives, in order, the indices of the vertices that we want to join. The first face is formed by joining the vertices $1, N + 1, N + 2$ and 2; the second face is formed by joining the vertices $2, N + 2, N + 3$ and 3; and so on. The `faces` matrix therefore must have the form:

$$
\begin{pmatrix}
1 & N+1 & N+2 & 2 \\
2 & N+2 & N+3 & 3 \\
\vdots & \vdots & \vdots & \vdots \\
N-1 & 2N-1 & 2N & N \\
N & 2N & N+1 & 1
\end{pmatrix}.
$$

The `faces` matrix can be defined as follows:

```
faces = [1:N; N+1:2*N; [N+2:2*N N+1]; [2:N 1]]';
```

We now have all the data we need to draw the patch:

```
clf
view(3)
h = patch('vertices',vert,...
'faces',faces,'facecolor','y');
xyz
```

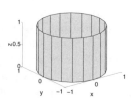

To colour the patch, we need to specify the temperature at each of the vertices. We assume a heat source is located at $(x, y, z) = (-.5, 0, 0.25)$, and that the temperature at any point on the cylinder is inversely proportional to its distance away from the source. The temperature at the vertices is calculated as follows:

```
dist = sqrt((x + 0.5).^2 + y.^2 + (z - 0.25).^2);
T = 1./dist;
```

We can now do the patch colouring:

```
colormap(hot)
set(h,'facevertexcdata',T',...
    'facecolor','interp',...
    'edgecolor','none');
```

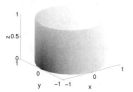

The colour of each vertex is indexed by the matrix T' to the colours in the hot colour map. (The hot spot is coloured dark in this gray scale version.) Each face of the patch is coloured in a way that interpolates between the colours of its vertices.

> **Exercise 22** *Can you see what we have done wrong in the above example? Hint: the patch colouring does not truly represent the distance away from the heat source. How would you go about getting a better representation? (Answer on page 193.)*

37.2 Light Objects

To create pictures representing real objects, you can colour them as if they are lit by one or more lights. The lights can be any colour you like, and the lit objects can have their reflectance properties adjusted to simulate different surfaces: mirror-like, or self-coloured and shiny, or dull. Lighting can be applied to surface or patch objects. Light objects themselves cannot be seen. For the following examples the `pltlight` function plots a dot at the position of each light on the graphic. Let us create a sphere and see what it looks like when lit:

```
clf
sphere
axis equal
grid,box,xyz
h = light('position',[1 -1 1]);
pltlight
```

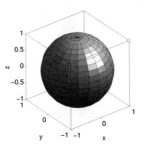

You should see a sphere with rather dull z-coded colouring and a glint of white light reflecting from about 45°N latitude. The dot at the top right hand corner of the plot is the result of the `pltlight` function, and represents the light. Let us see the result of using different coloured lights:

```
set(h,'color',[1 0 0])
light('position',[-1 1 1],...
    'color',[0 1 0])
light('position',[-1 -1 1],...
    'color',[0 0 1])
pltlight
```

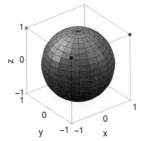

The light from the different coloured sources mix together to give a multicoloured shading. This graphic is still influenced by the z-coded colouring of the sphere itself. To show this more clearly, use the `flag` colour map:

```
colormap(flag)
```

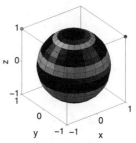

The flag colouring underlies the colour from the lights. To see the lights reflecting from a white sphere, set the colour of the surface to white. You could issue the statement `colormap([1 1 1])`, but the following achieves the same result:

```
h = findobj('type','surface');
set(h,'facecolor','w')
```

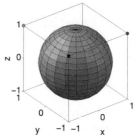

So far, we have been using the default "flat" method of rendering lit objects: each facet is a constant colour. But there are other ways. Let us generate a gray surface to work on:

```
clf
peaks(20),axis off
h = findobj('type','surface');
set(h,'FaceColor',[.5 .5 .5 ],...
    'edgecolor',[.5 .5 .5])
```

Generate a few lights:

```
x = [-3   -3    3];
y = [-3    3   -3];
z = [ 8    8    8];
```

```
cols = [1    1    1
        1    1    0
        0    1    0];
for i = 1:length(x)
 light('pos',[x(i),y(i),z(i)],...
       'color',cols(i,:))
end
pltlight
```

This is the flat lighting method. Other ways of rendering light are

`lighting gouraud`

`lighting phong`

The default is `lighting flat`.

Continuing the last example, we illustrate some different material properties:

`material metal`

`material shiny`

```
material dull
```

The default is `material default`.

The `material` and `lighting` commands are simple interfaces for changing the handle graphics properties that affect lighting. See the *Using MATLAB Graphics* manual for more details. For now, we illustrate some of the fine tuning effects that you can achieve. Here is the MATLAB logo, produced using the `membrane` function and illuminated:

```
clf
membrane
h = light('pos',[-.5 .5 .1]);
pltlight
axis off
```

The flat region near the light is a constant colour. That is because, by default, the light rays are parallel to the line joining the position of the light and the centre of the plot. To simulate a point source of light at the position of the light, set the `style` property of the light to `local`:

```
set(h,'style','local')
```

As another example, the following uses various reflectance properties of the (slightly roughened) sphere, along with a light off to one side, to produce a simulated crescent moon (see `moon.m` in companion software). First the spherical data points are randomized to produce a slightly rough sphere:

```
[x,y,z] = sphere(100);
N = size(x,1);
x = x + randn(N)/1000;
y = y + randn(N)/1000;
z = z + randn(N)/1000;
```

To get the following result I used trial and error to get the right values for the surface reflectance properties:

```
clf
h = surf(x,y,z);
set(h,'facecolor',[.5 .5 .5],...
      'edgecol','none')
hl = light('pos',[1000,0,0]);
axis equal
axis off
view(-20,0)
set(h,'specularstrength',0)
set(h,'ambientstrength',1)
set(h,'diffusestrength',10)
set(h,'backfacelighting','unlit')
```

38 MATLAB Programming

38.1 Vectorising Code

I once heard Cleve Moler say, "The `for` loop gets a bad rap." One of the clearest ways to see the truth of this statement will be in this section on speeding up MATLAB routines by vectorising code. We will be eliminating `for` loops and replacing them with operations on vectors or matrices. Yet, do not think that you must then eliminate all `for` loops from your MATLAB code. When `for` loops are used appropriately they are still very fast, efficient, and convenient. With this proviso, let us look at an example.

MATLAB's `diff` function takes the difference between successive pairs of elements in a vector, and writes them to another vector. Suppose you want to carry out a similar operation, except now you want to compute the sum of successive pairs instead of the difference. In mathematical notation, you would write the formula:

$$b_i = a_i + a_{i+1}, \qquad i = 1, 2, \ldots N - 1.$$

where a is the input vector of length N, and b is the output vector of pairwise sums. The following piece of MATLAB code would do the job:

```
N = length(a);
b = zeros(1,N - 1);
for i = 1:N-1
  b(i) = a(i) + a(i + 1);
end
```

This code, or at least the line inside the `for` loop, has the advantage of resembling the mathematical notation quite closely. We measure the time taken by this routine to calculate the pairwise sum of a 100,000-element vector (the code is saved in the m-file `forloop1` and we use `etime` to measure the execution time):

```
>>t0 = clock;forloop1;etime(clock,t0)
ans   =
     5.3545
```

A little over five seconds. Now we try to do it another way. Here is a diagram of what we want to do, with the elements we want to sum written out as indices of the respective vectors:

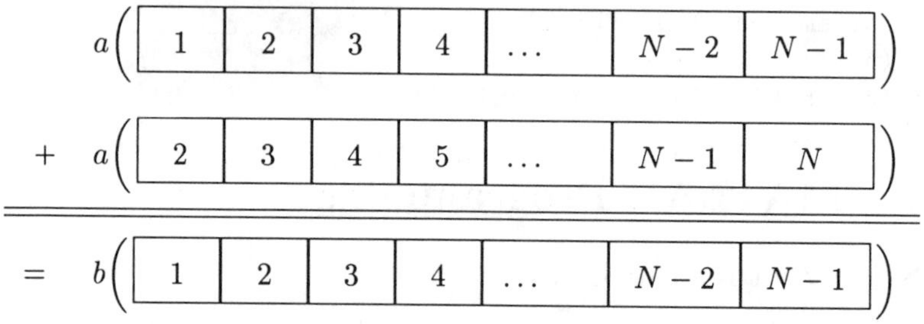

Writing the operation in this way allows us to see how we can use vectors of indices to do the summation. The top line of the sum can be written in MATLAB notation as `a(1:N-1)`, and the second line can be written as `a(2:N)`. The pairwise sum vector b therefore can be calculated in MATLAB with the following code:

```
b = a(1:end-1) + a(2:end);
```

We have used the special index `end` to refer to the final element of the vector. This time there is no advantage in pre-allocating the b vector as we did before the `for` loop above. With regular use, the vectorised MATLAB representation will seem to resemble the mathematical representation just as closely as the `for` loop. The time taken by this code is

```
t0 = clock;b = a(1:end-1) + a(2:end);etime(clock,t0)
ans   =
     2.2400
```

The vectorised version runs a little more than twice as fast as the `for` loop implementation.

Looping over matrix or vector subscripts can often be replaced by such matrix operations. Appropriate matrices can be generated using subscripting, as here, or by rearranging the input matrices using `reshape`, the transpose operator, or other matrix manipulations. MATLAB's columnar operations `sum`, `diff`, `prod`, etc. can then be used to do very fast operations on the appropriate matrices. MATLAB's suite of utility matrices can also come in handy when vectorising code (see `help elmat`).

We now look at a slightly more complicated example. Suppose you want to generate a matrix a_{ij} with elements:

$$a_{ij} = \begin{cases} \displaystyle\sum_{k=i}^{j} k & j \geq i \\[2ex] 0 & \text{otherwise} \end{cases}$$

The 5×5 version of a is

$$a = \begin{pmatrix} 1 & 1+2 & 1+2+3 & 1+2+3+4 & 1+2+3+4+5 \\ 0 & 2 & 2+3 & 2+3+4 & 2+3+4+5 \\ 0 & 0 & 3 & 3+4 & 3+4+5 \\ 0 & 0 & 0 & 4 & 4+5 \\ 0 & 0 & 0 & 0 & 5 \end{pmatrix}$$

$$= \begin{pmatrix} 1 & 3 & 6 & 10 & 15 \\ 0 & 2 & 5 & 9 & 14 \\ 0 & 0 & 3 & 7 & 12 \\ 0 & 0 & 0 & 4 & 9 \\ 0 & 0 & 0 & 0 & 5 \end{pmatrix}$$

A simple loop implementation of this calculation would resemble the following:

```
N = 200;
a = zeros(N,N);
for i = 1:N
  for j = 1:N
    if j>=i
      a(i,j) = sum(i:j);
    end
  end
end
```

Let us time this code (call it `forloop2`):

```
>>t0 = clock;forloop2;etime(clock,t0)
ans =
    2.9241
```

There are may different ways that we could vectorise this calculation, depending on our ingenuity. For now, we note that we can generate the first row of a by taking the columnar sum of the triangular matrix s:

$$
\texttt{a(1,:)} \;=\; \texttt{sum(s)}
$$

$$
= \;\; \text{sum}
\begin{pmatrix}
1 & 1 & 1 & 1 & 1 & \cdots \\
0 & 2 & 2 & 2 & 2 \\
0 & 0 & 3 & 3 & 3 \\
0 & 0 & 0 & 4 & 4 \\
0 & 0 & 0 & 0 & 5 \\
\vdots & & & & & \ddots
\end{pmatrix}.
$$

We can generate the second row of a by taking the same columnar sum but leaving out the first row of s:

$$
\texttt{a(2,:)} \;=\; \texttt{sum(s(2:end,:))}
$$

$$
= \;\; \text{sum}
\begin{pmatrix}
0 & 2 & 2 & 2 & 2 & \cdots \\
0 & 0 & 3 & 3 & 3 \\
0 & 0 & 0 & 4 & 4 \\
0 & 0 & 0 & 0 & 5 \\
\vdots & & & & & \ddots
\end{pmatrix}.
$$

In general, then, we can generate the ith row of a by taking the columnar sum of s leaving out its first $i - 1$ rows: `a(i,:) = sum(s(i:end,:))`. Our final code will consist of putting this statement inside a `for` loop (this will be a *good* use of a `for` loop—see the first paragraph in this section). Before we do that, though, we still need to generate the utility matrix s; here we can use matrix multiplication. The matrix we want can be obtained by taking the upper triangular part of the product of a column vector and a row vector:

$$
\begin{pmatrix}
1 & 1 & 1 & 1 & 1 & \cdots \\
0 & 2 & 2 & 2 & 2 \\
0 & 0 & 3 & 3 & 3 \\
0 & 0 & 0 & 4 & 4 \\
0 & 0 & 0 & 0 & 5 \\
\vdots & & & & & \ddots
\end{pmatrix}
= \text{triu}
\left[
\begin{pmatrix}
1 \\ 2 \\ 3 \\ 4 \\ 5 \\ \vdots
\end{pmatrix}
\cdot
\begin{pmatrix}
1 & 1 & 1 & 1 & 1 \cdots
\end{pmatrix}
\right]
$$

So here we have the final code to generate the a matrix (for $N = 200$):

```
N = 200;
s = triu((1:N)'*ones(1,N));
a = zeros(N,N);
for i = 1:N-1
   a(i,:) = sum(s(i:end,:));
end
a(N,:) = s(N,:);
```

The last row needs special treatment (see what happens when you let the loop run to i = N). On my computer this code took 2.1 seconds to execute, compared to 2.9 seconds for the simple for loop implementation given on page 177. We have saved nearly one second: not much, but if you have to repeat the calculation 10,000 times it becomes worthwhile.

38.2 M-File Subfunctions

MATLAB allows you to put more than one function in a file. If you put more than one function in a file, the second and subsequent functions are subfunctions; the first is the main function, or primary function. The idea is to have a file with the following structure:

```
function dinner = cook(entree,maincourse,dessert)
% Get matlab to cook a meal.
E = prepare(entree);
M = prepare(maincourse);
D = prepare(dessert);
dinner = [E M D];

function output = prepare(course)
switch iscourse(course)
  case 'entree'
    output = makeentree;
  case 'maincourse'
    output = makemaincourse;
  case 'dessert'
    output = makedessert;
  otherwise
    disp('Unknown course: do you really want to eat this?')
end
```

In this example prepare is the subfunction of the cook function. When MATLAB encounters the call to prepare, it checks to see if there is a subfunction called prepare in the same file before looking along the search path for an m-file called prepare. (Actually before looking along the path, it checks for the existence of a private subdirectory first. See the helpdesk if this intrigues you.) This means that you can give a subfunction the same name as an existing MATLAB function. The main function will use the subfunction and any other function will use the other existing function. As is true for single-file functions, subfunctions cannot "see" variables unless you pass them as arguments or declare them global. Subfunctions are invisible to help, which sees only the main function.

38.3   Debugging

MATLAB has a suite of debugging commands. A list of them can be obtained by typing `help debug`:

dbstop	Set breakpoint	dbstatus	List all breakpoints
dbclear	Remove breakpoint	dbstep	Execute one or more
dbcont	Resume execution		lines
dbdown	Change local	dbtype	List M-file with
	workspace context		line numbers
dbup	Change local	dbmex	Enable mex file
	workspace context		debugging
dbstack	List who called whom	dbquit	Quit debug mode

Other commands that are useful when debugging code are

keyboard Put this command in a function and the function will stop at that point and return control to the command window, but within the function's environment. This means that variables within the function can be accessed for printing out, plotting, etc. The command window prompt changes to K>> while the `keyboard` function is in effect. You can resume execution of the function by typing the character sequence r, e, t, u, r, and n at the K>> prompt,

echo Use the `echo` command to display each line of a script or function as it is executed.

diary The `diary` command is used when you want to save a copy of everything that appears in the command window, both what you type and what MATLAB types, in a file.

more The `more` command stops the screen from scrolling each time it fills with text. You can advance the screen one page at a time by pressing the space bar, or one line at a time by pressing the return key. If you press q when the screen is stopped, the current display will end at that point.

38.4   Profiler

The `profile` command measures the time taken to execute each line of code. Let us use it to examine the performance of the following code to produce an image of the Mandelbrot set (see companion software):

```
function mandelbrot
% MANDEL.M Produces a plot of the famous Mandelbrot set.
% see: http://eulero.ing.unibo.it/~strumia/Mand.html
% The generator is z = z^2+z0. Try changing the parameters:
N = 400;
```

```
xcentre = -0.6;
ycentre = 0;
L = 1.5;
x = linspace(xcentre - L,xcentre + L,N);
y = linspace(ycentre - L,ycentre + L,N);
[X,Y] = meshgrid(x,y);
Z = X + i*Y;
Z0 = Z;
for k = 1:50;
  Z = Z.^2 + Z0;
end
ind1 = find(isnan(Z));
ind2 = find(~isnan(Z));
Z(ind1) = 1;
Z(ind2) = 0;
contour(x,y,abs(Z),[.5 .5])
grid;box
axis equal off
```

You must tell the `profile` command which function you want to look at. The format of this command changed between MATLAB versions 5.2 and 5.3.

Profile in MATLAB 5.2

Initiate the profiler in MATLAB version 5.2 by typing:

`profile mandelbrot`

Now go ahead and run the function:

mandelbrot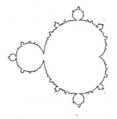

To see where MATLAB spent most of its time, type:

```
>> profile report
Total time in "mandelbrot.m": 30.12 seconds

100% of the total time was spent on lines:
      [15 21 12 18 17 19 11 20 22 16 ]
```

```
                    10: y = linspace(ycentre - L,ycentre + L,N);
0.13s,  0%    11: [X,Y] = meshgrid(x,y);
0.67s,  2%    12: Z = X + i*Y;
                    13: Z0 = Z;
                    14: for k = 1:50;
23.02s, 76%   15:     Z = Z.^2 + Z0;
0.02s,  0%    16: end
0.36s,  1%    17: ind1 = find(isnan(Z));
0.43s,  1%    18: ind2 = find(~isnan(Z));
0.22s,  1%    19: Z(ind1) = 1;
0.08s,  0%    20: Z(ind2) = 0;
5.15s, 17%    21: contour(x,y,abs(Z),[.5 .5])
0.02s,  0%    22: grid;box
                    23: axis equal off
```

Most of the time here is spent iterating the values of Z. You can get a plot of the time taken by the most time-consuming lines of code by capturing the output of the profile command and using it to produce a pareto chart:

```
>> t = profile
t =
         file: [ 1x64  char  ]
     interval: 0.0100
        count: [23x1  double]
        state: 'off'
>> pareto(t.count)
```

Here only the three most time-consuming lines (labelled on the x axis) are shown, the rest taking too little time to be of concern. The left-hand scale shows the time taken to do each line, in hundredths of a second. The line is the cumulative time. If we wanted to speed up this code, we would do well to concentrate on line 15, and forget trying to speed up the graphics.

Profile in MATLAB 5.3

The profile command has been significantly expanded in MATLAB 5.3. Use profile on to switch on the profiler. A hypertext report is produced by typing profile report. A graphical display of the profile results is obtained by typing profile plot.

39 Answers to Exercises (Part I, Basics of MATLAB)

Exercise 1 (Page 9)

The first three columns are a copy of the a matrix. The second three columns are the elements of a indexed by the elements of a. For example, a(a(3,2)) = a(8) = 6, which yields the marked element >6< of the answer:

```
>> [a a(a)]
ans =
     1     2     3     1     4     7
     4     5     6     2     5     8
     7     8     9     3    >6<    9
```

Exercise 2 (Page 39)

```
function out = posneg(in)

% Test for all positive (1), or all negative (-1) elements.

if all(in>0)
   out = 1;
elseif all(in<0)
   out = -1;
else
   out = 0;
end
```

Exercise 3 (Page 44)

The clown's hair is orange. You can use load clown to load the data (type clear first to get rid of any superfluous data). Typing whos will tell you that the workspace contains a matrix X and a variable map. Use image(X),colormap(map) to view the image.

Exercise 4 (Page 49)

We want to fit the data to an exponential curve:

$$p = Ae^{Bx} .$$

First we take logs to convert to a linear equation:

$$\log p = \log A + Bx .$$

We change variables:

$$p' = C + Bx.$$

Now we simply do a least-squares fit using this equation; that is, a straight line. We could use the backslash notation that we used to fit the parabola but, for variety, let's use the `polyfit` function. A straight line is a polynomial of degree 1. The following code takes the logarithm of the population data and fits a straight line to it:

```
>> logp = log(P);
>> c = polyfit(year,logp,1)
c =
      0.0430  -68.2191
```

The vector c contains B and C, in that order. We use the polynomial evaluation function `polyval` to calculate the fitted population over a fine year grid:

```
>> year_fine = (year(1):0.5:year(length(year)))';
>> logpfit = polyval(c,year_fine);
```

And we display the results on linear and logarithmic y-scales:

```
subplot(221)
plot(year,P,':o',year_fine,exp(logpfit))
subplot(222)
semilogy(year,P,':o',year_fine,exp(logpfit))
```

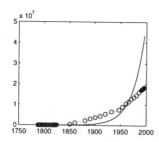 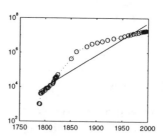

The single straight line cannot fit all the data. The right hand plot indicates that there were two growth factors, B: one prior to 1870 and one after. Let's do another fit using only the data after 1870:

```
ind = find(year>1870);
logp = log(P(ind));
c = polyfit(year(ind),logp,1);
logpfit = polyval(c,year_fine);
clf subplot(221)
plot(year,P,':o',year_fine,exp(logpfit))
subplot(222)
semilogy(year,P,':o',year_fine,exp(logpfit))
```

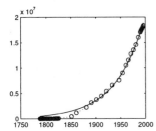 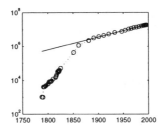

If you zoom in on the right hand plot you'll find that this growth rate is too fast for the period between 1990 and 1996.

Exercise 5 (Page 55)

The following m-file illustrates how to generate a 2-dimensional sinusoid and its FFT. Experiment with the relative x and y frequencies and see the effect on the FFT. Try different functions of x and y. Try adding some noise. Try plotting the logarithm of P.

```
t=linspace(-pi,pi,64);
[x,y]=meshgrid(t);
z = sin(3*x + 9*y);
Z = fft2(z);
P = fftshift(abs(Z).^2);
f = fftfreq(0.5, length(t));

clf
colormap([0 0 0])
subplot(221)
mesh(x,y,z)
axis([-pi pi ...
      -pi pi ...
      -15 15])
view([60 50])
xlabel('x')
ylabel('y')
title('Signal')

subplot(223)
mesh(f,f,P)
axis tight
view([60 50])
xlabel('x-frequency')
ylabel('y-frequency')
title('Transform')
```

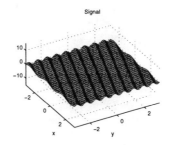

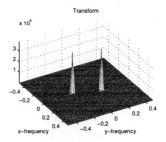

Exercise 6 (Page 59)

You can generate a sinusoidal frequency variation by specifying a sinusoid input to the voltage controlled oscillator function vco. Get the results of specgram by using output arguments and compare a plot of the results using an image and a surface plot:

```
Fs = 1000;
t = linspace(0,2*pi,8192);
x = sin(t);
y = vco(x,[0 500],Fs);
[z,freq,time] = specgram(y,[],Fs);
p = 20*log10(abs(z));
subplot(221)
  imagesc(time,freq,p)
  axis xy
  colormap(flipud(gray))
  colorbar
  xlabel('Time, sec')
  ylabel('Frequency, Hz')

subplot(223)
  surfl(time,freq,p)
  shading flat
  xlabel('Time, sec')
  ylabel('Frequency, Hz')
  zlabel('Power, dB')
```

40 Answers to Exercises
(Part II, Beyond the Basics)

Exercise 7 (Page 74)

To repeat the calculation for the case of 100 nodes, we do the following:

```
dt = 2*pi/100;
t = dt:dt:100*dt;
x = cos(t)';y = sin(t)';
xy = [x y];
e = ones(100,1);
A = spdiags(e,2 ,100,100) + ...
    spdiags(e,50,100,100) + ...
    spdiags(e,98,100,100);
A = A +
A';
subplot(221)
spy(A)
subplot(222)
gplot(A,xy)
axis equal off
```

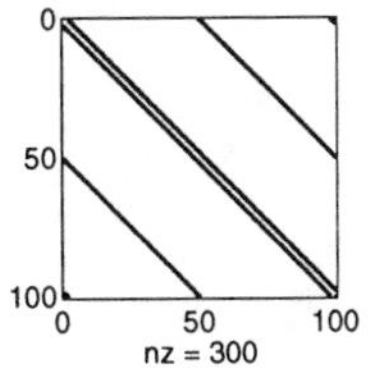

The next part of the exercise is to change the connection matrix. An interesting one is the geometrically parallel network:

```
A = spdiags(e, 25,100,100) + ...
    spdiags(e,-75,100,100);
A = fliplr(A);
subplot(221)
spy(A)
subplot(222)
gplot(A,xy)
axis equal off
```

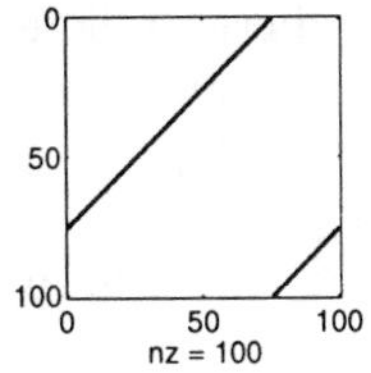

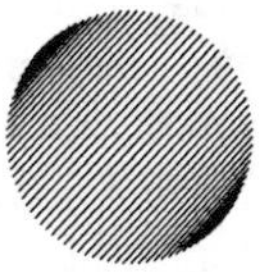

Exercise 8 (Page 75)

This will produce a list of the ASCII characters corresponding to the integers from zero to 255:

```
I = (0:255)';
[int2str(I) blanks(256)' char(I)]
```

Some of the output is shown below:

33 !	36 \$	39 '	42 *	45 –	48 0
34 "	37 %	40 (	43 +	46 .	49 1
35 #	38 &	41)	44 ,	47 /	50 2

Typing char(7) rings the bell.

Exercise 9 (Page 80)

The strvcat function is used instead of char because it ignores empty strings in the input; the char function doesn't:

```
>> char('','The','','quick')
ans =

The

quick
>> strvcat('','The','','quick')
ans =
The
quick
```

If char were used instead of strvcat, the result would always begin with a blank line.

Exercise 10 (Page 83)

The problem is to deal with the two cases: (1) where the name of a function or m-file is given, such as 'sin', and (2) where the function itself is given, such as 'sin(x)'. The difference here boils down to whether the string input contains brackets or not (see hint). In other cases the string input might not contain brackets, but would contain characters used in defining a function, such as +, -, *, /, or . (as in t.^2). The ASCII values for these characters are all less than 48, so we detect the presence of a function (rather than a function *name*) by checking the input string for ASCII values less than 48. If this is the case, we make the input string into an inline function before passing it to feval:

```
function funplot(f,lims)

% Simple function plotter.

% Test for characters whose presence would imply that f
% is a function (not a function name):
if any(f<48)
   f = inline(f);
end
```

```
x = linspace(lims(1),lims(2));
y = feval(f,x);
clf
plot(x,y)
```

(This trick is used in the m-file `fplot` which is a more elaborate version of our `funplot`. `fplot` adapts the plotting grid to the local behaviour of the function, putting in more points where the gradient is steep.)

Exercise 11 (Page 86)

The "stuck" in question is indicated by the arrow in the following plot:

```
>> t = {'help' spiral(3) ; ...
        eye(2) 'I''m stuck'};
>> tt = {t t ;t' fliplr(t)};
>> tt{2,2}{2,1}(5:9)
ans =
stuck
>> cellplot(tt)
```

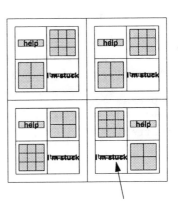

Exercise 12 (Page 94)

The difference between `meshgrid` and `ndgrid` for less than four input arguments is that the first two output arguments are transposed. This makes it convenient to do x-y plots using the [x,y] outputs of the `meshgrid` command. The outputs of the `ndgrid` command follow the logical ordering of indices in MATLAB: if [u,v,w] = ndgrid(...) then u's elements will vary over its rows, v's elements will vary over its columns, and w's elements will vary over its pages.

Exercise 13 (Page 97)

The distance d of each point from (x_0, y_0) is given by:

$$d = \sqrt{(x - x_0)^2 + (y - y_0)^2},$$

so we calculate this for the centres of the red, green, and blue regions. Then we find the points outside the radius and set them equal to zero:

```
R = 1; N = 200;
[x,y] = meshgrid(linspace(-2,2,N));
r = sqrt((x + 0.4).^2 + (y + 0.4).^2);
ind = find(r>R);
r(ind) = 0;
g = sqrt((x - 0.4).^2 + (y + 0.4).^2);
ind = find(g>R);
g(ind) = 0;
b = sqrt(x.^2 + (y - 0.4).^2);
ind = find(b>R);
b(ind) = 0;
rgb = cat(3,r,g,b);
imagesc(rgb)
axis equal off
```

You may find that the image on the screen has been dithered because we now have a very large number of colours. You might like to investigate other combinations of red, green, and blue matrices, for example:

```
r = peaks;
g = r';
b = fftshift(r);
```

Mixing linear ramps of colour in different directions is interesting. Be adventurous!

Exercise 14 (Page 115)

Since we know the dimensions of the borders in pixels, we set the figure's units property to pixels. It is then just a matter of taking into acount the correct number of borders when defining the figures' positions which, remember, don't include the borders. The following code does the job. We start by getting the size of the screen (the root object, whose handle is alway zero) in pixels.

```
set(0,'Units','pixels')
screensize    = get(0,'ScreenSize');
edgethickness = 5;
topthickness  = 10;
scr_width  = screensize(3);
scr_height = screensize(4);
figwidth   = scr_width/2  - 2*edgethickness;
figheight  = scr_height/2 - ...
        2*(edgethickness + topthickness);
pos1 = [edgethickness,...
        edgethickness,...
        figwidth,...
        figheight];
```

```
pos2 = [scr_width/2 + edgethickness,...
        edgethickness,...
        figwidth,...
        figheight];
pos3 = [scr_width/2  + edgethickness,...
        scr_height/2 + edgethickness,...
        figwidth,...
        figheight];
figure('Position',pos1)
figure('Position',pos2)
figure('Position',pos3)
```

The width of the window border might be different from these on your computer. There is no way of obtaining these widths from within MAT-LAB. You might have to resort to trial and error to get the window thicknesses exactly right for your computer.

Exercise 15 (Page 118)

Issue the command **type gcf** and you will see that if there are no figures, gcf creates one, whereas get(0, 'CurrentFigure') doesn't.

Exercise 16 (Page 119)

Did you have fun?

Exercise 17 (Page 123)

The following commands should produce the required display. First we generate a grid of 100×100 points over the interval $[0, 1]$:

```
N = 100;
v = linspace(0,1,N);
[x,y] = meshgrid(v);
```

We want to draw a vertical line at each of the grid points to represent the vines of the vineyard or trees of the orchard. We string out the x and y grid points into two row vectors and use matrix multiplication to duplicate these. The z values, representing the start and end points, go from zero to a height of 0.01:

```
x = [1; 1]*x(:)';
y = [1; 1]*y(:)';
z = [zeros(1,N^2); 0.01*ones(1,N^2)];
plot3(x,y,z,'r')
```

(The plot might take a few seconds to render; be patient.) We want to modify this plot so that a perspective projection is used, and with a viewpoint as if we were standing near the edge of the vineyard:

```
set(gca,'proj','per')
axis equal
set(gca,'cameraposition',[.5 -1 .2])
axis vis3d off
set(gca,'cameraposition',[.5 -.1 0.03])
```

Exercise 18 (Page 132)

You should not use `title` instead of `text` because the title disappears when you do `axis off`.

Exercise 19 (Page 145)

The following code does the job. Items are "grayed out" by setting their "enable" property to "off".

```
uimenu('Label','File')
uimenu('Label','View');
E = uimenu('Label','Edit')
uimenu('Label','Options')
uimenu(E,'Label','Cut','Enable','off')
uimenu(E,'Label','Copy','Enable','off')
uimenu(E,'Label','Paste')
uimenu(E,'Label','Rotate','Separator','on')
S = uimenu(E,'Label','Scale')
uimenu(S,'Label','10%',...
          'Enable','off')
uimenu(S,'Label','50%',...
          'Enable','off')
uimenu(S,'Label','150%',...
          'Enable','off')
uimenu(S,'Label','200%',...
          'Enable','off')
uimenu(S,'Label',...
          'Custom Scaling...')
```

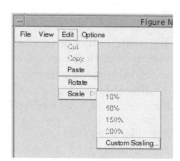

Exercise 21 (Page 164)

A truncated pyramid can be produced using the following code:

```
x = [0  0  1  1  0 .4
     1  1  1  0  0 .4
     1 .6 .6 .4 .4 .6
     0 .4 .6 .6 .4 .6];
y = [0  1  1  0  0 .4
     0  1  0  0  1 .6
     1 .6 .4 .4 .6 .6
     1 .6 .6 .4 .4 .4];
z = [0  0  0  0  0  1
     0  0  0  0  0  1
     0  1  1  1  1  1
     0  1  1  1  1  1];
patch(x,y,z,'y')
view(3);xyz,box
```

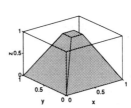

Exercise 22 (Page 171)

The faces are coloured according to the colours of the vertex. We have
vertices defined for $z = 0$ and $z = 1$, but the heat source is located at
$z = 0.25$. The result should be symmetric about the $z = 0.25$ plane,
but our result does not have this symmetry. The reason is that we only
have two z values. To produce a better display we simply need to add
vertices at a range of z values. The following code does the job:

```
N = 100; % Number of points around the circle
M = 30; % Number of circles in the cylinder
dt = 2*pi/N;
t = (0:dt:(N-1)*dt)';
h = linspace(0,1,M); % vector of heights
xv = cos(t);
yv = sin(t);

% Reproduce the vertices at different heights:
x = repmat(xv,M,1);
y = repmat(yv,M,1);
z = ones(N,1)*h;
z = z(:);
vert = [x y z];

% These are the facets of a single 'layer':
facets = zeros(N,4);
facets(1:N-1,1) = (1:N-1)';
facets(1:N-1,2) = ((N+1):(2*N-1))';
facets(1:N-1,3) = ((N+2):(2*N))';
facets(1:N-1,4) = (2:N)';
facets(N,:) = [N 2*N N+1 1];
```

```
% Reproduce the layers at the different heights:
faces = zeros((M-1)*N,4);
for i=1:M-1
  rows = (1:N) + (i - 1)*N;
  faces(rows,:) = facets + (i - 1)*N;
end

%Define heat source and temperature:
xs = -0.5;
ys = 0;
zs = 0.25;
dist = sqrt((x - xs).^2 + (y - ys).^2 + (z - zs).^2);
T = 1./dist;

clf
colormap(hot)
h = patch('vertices',vert,'faces',faces,...
    'facevertexcdata',T,...
    'facecolor','interp',...
    'linestyle','none');
view(78,36)
axis equal
% Plot the source:
hold on
plot3([xs xs],[ys ys],[0 zs])
plot3(xs,ys,zs,'*',...
    'markerSize',12)
```

In the resulting graphic the vertices are shown as points and the source is shown as the star on the stick.

Index